The Interpersonal-Psychological Theory of Attempted and Completed Suicide

Conceptual and Empirical Issues

Freddy A. Paniagua, Sandra A. Black,
M. Shayne Gallaway and Michelle A. Coombs

authorHOUSE®

AuthorHouse™
1663 Liberty Drive
Bloomington, IN 47403
www.authorhouse.com
Phone: 1-800-839-8640

First published by AuthorHouse   10/19/2010

ISBN: 978-1-4520-8152-6 (e)
ISBN: 978-1-4520-8153-3 (sc)

Library of Congress Control Number: 2010914362

Printed in the United States of America

This book is printed on acid-free paper.

CONTENTS

AKNOWLEDGEMENTS

We want to thank Professor David Lester, Richard Stockton College of New Jersey, for his encouragement and editorial supports.

The production of this book was supported in part by an appointment (Dr. Freddy A. Paniagua) to the Faculty Research Participation Program at the U. S. Army Public Health Command (USAPHC; formerly U.S. Army Center for Health Promotion and Preventive Medicine) administered by the Oak Ridge Institute for Science and Education through an interagency agreement between the U.S. Department of Energy and USAPHC.

The authors want to affirm that the views expressed herein are the views of the authors and do not reflect the official policy of the U.S. Department of the Army, the U.S. Department of Defense, the U.S. Department of Energy, the Oak Ridge Institute for Science and Education, or the U.S. government.

PREFACE

The motivation for writing this book has a long history. Professor Thomas Joiner published his theory of suicide in 2005 in a book entitled, "Why People Die by Suicide," Harvard University Press. In 2005, Dr. Paniagua was a Tenured Professor in the Department of Psychiatry and Behavioral Sciences, University of Texas Medical Branch (UTMB), Galveston. He read Dr. Joiner's book in 2006, and since that time he was intrigued by two monumental assertions in Dr. Joiner's book.

First, according to Dr. Joiner's theory, "by employing three simple concepts [i.e., thwarted belongingness, perceived burdensomeness, and the ability to enact lethal self-injury]" (Joiner, 2005, p. 226), the theory is able to explain suicide attempts and "all deaths by suicide worldwide, across cultures" (p. 226). As a result of Dr. Paniagua's familiarity with theories of suicide at that time, it was clear to him that Dr. Joiner had made a monumental assertion because none of the alternative theories of suicide available in 2005 made any similar assertion (Lester, 2004a; Lester, 2008). The reason for this observation is that suicidal acts (suicide attempts and completed suicides) are among the most complex human situations requiring multiple explanations (i.e., multiple factors are always involved when someone decides to take his/her own life).

The second monumental assertion that intrigued Dr. Paniagua was the use of the theory to explain why certain victims of 9/11 jumped to their death from the upper floors of the World Trade Center (WTC). According to Dr. Joiner's theory, these victims elected to kill themselves; they were not actually killed by al-Qaida! In the same context, Dr. Joiner's theory also claims that the 9/11 al-Qaida terrorists died by suicide and not by an act of martyrdom.

Dr. Paniagua kept these two assertions in mind for several years (2006-2010), with the hope that one day he would have time to write a peer-reviewed article to explain his thoughts regarding the core assumption in Dr. Joiner's (2005) theory (the first monumental assertion) and particularly, his disagreement with Dr. Joiner's theory regarding the explanation of some victims of 9/11 as suicides rather than homicides resulting from the al-Qaida terrorist attacks (the second monumental assertion).

In 2009, Dr. Paniagua retired from UTMB, and was encouraged by Dr. Sandra A. Black to take a position at Aberdeen Providing Ground, Maryland, to help with the development of the *Behavioral and Social Health Outcomes Program* (BSHOP). This program was specifically established to conduct surveys, focus groups, surveillance, and other activities to understand the rate of suicide attempts and completed suicides in the U.S. Army. This program (BSHOP) is associated with the Faculty Research Participation Program at the U. S. Army Public Health Command (USAPHC; formerly U.S. Army Center for Health Promotion and Preventive Medicine) administered by the Oak Ridge Institute for Science and Education (ORISE) through an interagency agreement between the U.S. Department of Energy and USAPHC.

The Faculty Research Participation Program administered by ORISE is designed to provide participants with practical experience and help them with the development of career aspiration (particularly in the case of scholars with substantial contributions in the behavioral sciences, but with interest to explore new scientific possibilities). This is an excellent program, and interested individuals in any scientific area should contact: www.orise.orau.gov.

Dr. Paniagua agreed to join BSHOP, through the auspices of ORISE, for three reasons. First, he had been very concerned about the rates of suicide in the Army long before he was invited to apply to the ORISE opportunity (Paniagua, 2009). Second, the program provided Dr. Paniagua with solid training in the epidemiology of suicidal behavior in the Army. Third, the program provided him with an understanding of the complexity of suicidal behavior in the Army, in contrast to the civilian population. In addition, ORISE was a significant factor in Dr. Paniagua's contribution to the literature involving suicide attempts and completed suicides in the Army (Paniagua, 2010a).

Lieutenant Colonel Michael R. Bell, M.D., MPH, the Program Manager of BSHOP at the time Dr. Paniagua joined this program, suggested a review of existing theories of suicide and how these theories could be used as a theoretical background for the findings of BSHOP's surveillance and investigative suicidal activities that were being headed by Dr. Black. This new position and Dr. Bell's suggestion provided the opportunity to return to Dr. Joiner's theory to further explore the aforementioned monumental assertions.

Despite the busy schedule at BSHOP, including extensive investigations of Army suicides and traveling to various Army installations to conduct surveys and focus groups with Soldiers, Dr. Paniagua found time to work on Dr. Joiner's theory. At that time, he enlisted the three co-authors to help in the preparation of a peer-reviewed article. Dr. Sandra A. Black, Dr. M. Shayne Gallaway, Ms. Michelle A. Coombs, and Dr. Paniagua met numerous times and conducted a thorough literature review of the theory, including reviewing empirical peer-reviewed articles that supported the core assumptions of the theory.

During that literature review (covering the years 2006-2010), no single article was found that questioned the core assumptions of the theory, particularly the second monumental assertion. This suggested that something was wrong, in that either leading theoretical experts on suicide (e.g., Lester, 2008) were more interested in dealing with theories of suicide that are multifactorial (which is the opposite of Dr. Joiner's theory) or that they did not want to spend time with a theory that is essentially reductionist in terms of claiming to be the final theoretical explanation of all suicide acts.

In addition, it appears that those theoretical suicide experts may not have taken the opportunity to carefully screen Dr. Joiner's theory. For example, Drs. Paniagua and Black met with Dr. John J. Mann (from Columbia University, New York) on November 17, 2009 (Dr. Mann is among the leading experts in the present context, e.g., Mann, Waternaux, Hass, & Malone, 1999), and during the meeting Dr. Paniagua discussed the second monumental assertion in Dr. Joiner's theory. Dr. Mann was surprised and completely disagreed with that assertion.

In terms of the above scenario, we knew that we were on the right track writing a critique of Dr. Joiner's theory, with the hope that our critique would be accepted in the peer-reviewed process.

Because of the theoretical nature of our critique, we selected *Psychological Review* as our first venue to publish the paper dealing with several critical issues in Dr. Joiner's (2005) theory of suicide. This journal is published by the American Psychological Association and it emphasizes the publication of theoretical papers. We submitted the paper on January 27, 2010. According to the editor (Dr. John R. Anderson, Carnegie Mellon University, Pittsburgh, Pennsylvania) we would have to wait from 60 to

90 days before receiving a final decision from him regarding the status of the paper. The paper was assigned to Dr. Charles S. Carver (University of Miami, Coral Gables, Florida), one of the associate editors for *Psychological Review*. Four days later (January 31, 2010), however, we were surprised to receive an email from Dr. Carver explaining his reason for "declining [the paper] without outside reviewers" in the peer-review process established by *Psychological Review*.

Readers may think Dr. Carver's decision was not fair. Actually, his decision was within standard procedures for peer-reviewed journals, in which either the editor or associate editor often makes a decision regarding the status of a paper without the need to send it for external review. For example, a manuscript may not be sent for external peer-review because its content is not appropriate for the journal, it is incomplete or missing details necessary to understand its rationale and methodology, or the paper has some merit but a similar paper is currently under review or has been already accepted by the editor. In the latter case, the editor or associate editor may recommend the author prepare a commentary of the paper under review or accepted by the journal (which was the case with our paper, as we explain below).

The good news in our communication with Dr. Carver is that he actually believed that our paper would make a contribution to the suicidology literature. In his email, Dr. Carver wrote: "*Your argument is at its most persuasive when you point to omissions in literature cited and characterized by Joiner. If he* [Dr. Joiner] *has presented a misleading picture of the existing literature, that does indeed constitute a problem. Similarly, if none of the studies that have been used to support the theory are actually testing the theory, that is also a problem*" (italics added).

Dr. Carver, however, decided our paper could not be sent out for external review because "*A complicating factor here is that a manuscript by Joiner and colleagues detailing the theory in its current incarnation has recently been accepted (by the preceding editorial team) for publication in this journal. I have not read it, and I do not know how much it differs from previous statements. If this were not the case, I would recommend that you revise (and shorten and sharpen) the paper and submit it to a more specialized journal. Indeed, that may be the best option in any case*" (italics added).

In the same email Dr. Carver changed his earlier decision and wrote: "However, *given that Psych Review will be publishing an article presenting a theory that you believe is seriously flawed, an argument could be made that a commentary on its flaws would be appropriate to the journal. For that reason,*

I am willing to entertain the possibility of your submitting a revision, under the following circumstances. First, please contact Dr. Joiner to obtain a preprint of the accepted article, so that you can determine the extent of overlap with and extension from, previous articles. Second, the revision should focus most explicitly on the most recent formulation and portrayal of support. If you believe previous articles mischaracterized evidence, but those mischaracterizations are not included in the new one, you may mention them, but don't dwell on them. Third, I would recommend you spend less attention on specific comparisons with previous theory…Focus on how the theory is inadequate to cover the territory and on problems in the tests of the theory" (italics added).

We decided, however, to follow Dr. Carver's recommendation to *"revise … the paper and submit it to a more specialized journal"* (italics added) for the following reasons. First, we felt that asking us to directly contact Dr. Joiner was not among the expected editorial standards in peer-reviewed journals. If the editor or associate editor feels that a commentary would be needed to clarify issues in the accepted paper, the editor or associate editor would contact the senior author of the paper that needs a commentary and then establish the process for a communication between the senior author and the individual who would submit that commentary to the peer-reviewed process. Second, Dr. Carver 's suggestion to write a commentary on Dr. Joiner's article accepted by Psychological Review would have involved eliminating many details we felt were necessary to address critical issues in Dr. Joiner's (2005) theory.

Dr. Paniagua then conducted a review of possible "specialized journals" (i.e., journals dealing with the topic of this book), and selected *OMEGA: Journal of Death and Dying* (suggested to us by Professor David Lester, see below), for two reasons. First, a response to an email sent to the editor on February 2, 2010 (Dr. Kenneth J. Doka, The College of New Rochelle, New York) indicated our paper was very appropriate for OMEGA. Second, in that response we learned the journal has "no limit per se [regarding the number of pages submitted]." Thus, the manuscript was submitted to OMEGA on February 5, 2010.

We received a decision letter on August 9, 2010 that was somewhat confusing.

Dr. Doka first stated that he wished "to thank you for sending your article. Unfortunately, it is not suitable for *Omega*." Then, Dr. Doka ended his letter with the following statement: *"You may wish to resubmit if you can address the issues noted by reviewers"* (italics added). It appeared that Dr. Doka was also confronted with a dilemma similar to that encountered by Professor Carver, namely, rejecting a paper that appears to be on the right track in terms of a critique of Dr. Joiner's theory. In addition, it was not clear what we were expected to do based on *OMEGA*'s reviewers' comments. Generally speaking, all comments were positive. For example, reviewers wrote that *"the main tenet of the model* [i.e., Dr. Joiner's theory] *is that actual suicide attempts and completions can be accounted for by a 3-way interaction between perceptions of burdening others, social alienation, and the capacity to commit suicide. Only when all of these elements are present will people attempt and/or completed suicide. In an empirical reported cited by the author, Joiner et al. (2009) note that there have been no bonafide tests of his central thesis, only tests of various elements"* (italics added). This was, precisely, the core argument we made in our paper submitted to *OMEGA* when discussing violations of fundamental assumptions in Dr. Joiner's theory (extensively discussed in the present book).

Furthermore, OMEGA's reviewers pointed out that the *"most convincing element of the authors' argument is that it is very difficult to test Joiner's 3-way interaction…without considering psychopathology, which Joiner acknowledges is present in practically everyone who attempts or completes suicide "* (italics added). This is another point discussed extensively in this book.

Finally, one of the reviewer for OMEGA pointed out that *"as the authors appear to believe, I find Joiner's model to be simplistic and reductionist, but I am happy to see that his work has stimulated some interest in suicidology"* (italics added). We were pleased to learn that this reviewer agreed with us in terms that Dr. Joiner's theory is a reductionistic theory in nature. We also agree with *OMEGA*'s reviewers in that Dr. Joiner's theory has stimulated *"some interest in suicidology"* (italics added), which is also a point we make in this book (see the Conclusion).

As a result of these reviews, Dr. Paniagua felt that it would be very difficult to publish our paper via the standard peer-review process, and made the decision to submit the paper as a book to *AuthorHouse*®. It appeared to us that our critiques of Dr. Joiner's (2005) theory have some merit in the peer-review process. At the same time, however, we thought that it would be a difficult and lengthy endeavor to make our critiques of Joiner's (2005) theory available to the scientific community via the standard peer-review process. *AuthorHouse*® is not a peer-review publishing company, but it is a good choice when one thinks that critical arguments not accepted by the scientific editorial/peer-review process need to be published.

Dr. Paniagua's decision was also influenced by a note he received from Dr. David Lester (who had been cited by Dr. Joiner in the original publication of the theory in 2005). Dr. Lester is considered among the most influential theoreticians in the field of suicidology (e.g., Lester, 1990; Lester 1994; Lester 2005a; Lester 2004b; Lester, 2008), and is a Distinguished Professor of Psychology at the Richard Stockton College of New Jersey. On November 16, 2009, Dr. Paniagua sent Dr. Lester a copy of the manuscript and informed him about the plan to submit it to Psychological Review. On November 21, 2009, Dr. Lester sent a response to Dr. Paniagua in which he stated: *"I read your paper on Joiner's theory. It is a very thorough work. It reminded me of my analysis and critique of Henry and Short's theory which I did back in the 1970s. People rarely go into theory in such detail-pointing out the errors and inconsistencies. I hope you get it published ...if Psychological Review declines to publish it, where do you plan to submit it next?"* (italics added). Professor Lester went on to suggest three additional peer-reviewed journals for the submission of our paper: *Archives of Suicide Research-ASR, Suicide and Life-Threatening Behavior-SLTB* (edited by Dr. Joiner), and *OMEGA*. Professor Lester wrote, *"I wonder if* [Dr. Joiner] *would consider it."* For obvious reasons, Dr. Paniagua decided not to send the paper to SLTB, and the total number of pages was probably too high for ASR. OMEGA was our best choice, for reasons explained above.

In the end, Dr. Paniagua felt that the best choice was to submit the paper to *AuthorHouse*®, with the assumption that readers of this book would have the opportunity to determine whether or not it is actually making valid points.

INTRODUCTION

The World Health Organization-WHO (2009) has designated suicide a major public health problem, nationally and internationally. According to WHO (2009), every year about one million people die by suicide, and the worldwide rate is approximately 16 per 100,000 (approximately one death by suicide every 40 seconds). Over the past 45 years, suicide rates have increased by 60% worldwide, and suicide is considered among the three leading causes of death among those aged 15-44 years in many countries, and the second leading cause of death in 10-24 year age group.

Worldwide, suicide attempts are 20 times more frequent than completed suicides. In the U.S, about 30,000 individuals elect to die by suicide each year, translating to more than 80 deaths by suicide per day. Suicide is the 11th leading cause of death in the U.S., and the second leading cause of death among those aged 25-34 years (CDC, 2009a). Everyone is at risk for suicide, regardless of age, gender, professional affiliation, marital status, race, religious affiliation, or socio-economic status.

Suicide rates, however, vary consistently across some demographic characteristics. For example, in 1897 Emile Durkheim concluded that suicide "happens to be an essentially male phenomenon" (Durkheim, 1897/1951, p. 72) and 113 years later the rate of deaths by suicide among men still exceeds those among women (Gold, 2006). In addition, suicide rates are generally higher among elderly males (CDC, 2009b), and among people with diagnoses of mental disorders (Simon, 2006).

Because of the negative impact of suicide on all segments of society (CDC, 2009a; CDC, 2009b; Gold, 2006; Simon, 2006; WHO, 2009), the need to understand why some individuals attempt or elect to die by suicide has been a clinical and academic concern for many years, at least since the publication of the first widely recognized theory of suicide by Durkheim (1897/1951). Currently, the number of theories of suicide is over

15 (Ellis, 2007; Heeringen, 2001a; Kerkhof & Arensman, 2001; Leenaars, 2008; Lester, 1994; Lester, 2004a; Lester, 2004b; Lester, 2008; Mann, Waternaux, Haas, & Malone, 1999).

The multiplicity of theories of suicide is explained by the multifactorial nature of suicide (Lester, 1990; Simon, 2006). Despite the good intentions of the authors of such theories, none of the existing theories of suicide has been able to satisfactorily explain why people attempt suicide or die by suicide. This has been a historical fact in the field of suicidology, until the publication of the *Interpersonal-Psychological Theory of Attempted and Completed Suicide* (IPT-ACS) by Joiner (2005). For example, in the original formulation of the theory Joiner (2005) recognized the contribution of existing theories of suicide (see Joiner, 2005, pp. 32-45), but he claimed that his theory is the one that offers conclusive theoretical explanations regarding why people either attempt suicide or die by suicide. According to Joiner (2005), "by employing three simple concepts [i.e., thwarted belongingness, perceived burdensomeness, and the acquired ability to enact lethal self-injury]" (p. 226) the IPT-ACS is able to explain suicide attempts and *"all deaths* by *suicide worldwide, across cultures"* (p. 226, italics added).

Therefore, Joiner's theory is unique because it claims the *final and unquestionable explanation* of suicide acts (suicide attempts and completed suicides) worldwide with the help of only three constructs (described below; see Joiner, 2005, p. 226). The unique feature of Joiner's theory (2005) also points to a theory that is essentially reductionist in nature (Paniagua, 2010b). The reason for this conclusion, as many authors have recognized, is that "suicide is a multifaceted event that is open to study from multiple points of view [i.e., theories]" (Leenaars, 2008, p. 15; see also Simon, 2006). Among the theories Joiner (2005) used to support the need for his theory, Durkheim's theory (described below) clearly meets the metric of reductionism. As noted by Lester (2008), pertaining to Durkheim's theory, only two constructs are necessary to explain why people either attempt suicide or die by suicide: the level of social integration and the level of social regulation. Joiner's theory shares this metric of reductionism, using only three constructs in the suicidal equation to explain why people die by suicide: thwarted belongingness, perceived burdensomeness, and the acquired ability to enact lethal self-injury. Shneidman's (1987) cubic theory of suicide (discussed below) is also reductionist in nature, because he tried to explain suicidal behavior with only three constructs (i.e., psychache, psychological pressure, and perturbation), and "insisted that every person who commits suicide is psychologically in [that] cubelet at the time of the act" (Jobes & Nelson, 2006, p. 35). Shneidman (1996), however, later proposed a theory of suicide that was more multifactorial.

[Joiner's theory is known by three titles: *The Interpersonal-Psychological Theory of Suicide* (Joiner & Van Orden, 2008), *Interpersonal Theory of Suicide* (without the "psychological" term; Joiner, Van Orden, Witte, & Rudd, 2009), and *The Interpersonal-Psychological Theory of Attempted and Completed Suicide* (IPT-ACS). The last title most closely reflects the central cores of the theory, and this is the title selected for this book; see Wingate, Burns, Gordon, Perez, Walker, Williams, & Joiner, 2006, p. 272.]

Since the publication of the IPT-ACS in 2005, a number of conceptual papers and empirical studies have appeared in the suicidology literature further elaborating on different aspects of the IPT-ACS (e.g., Anestis, Bryan, Cornette, & Joiner, 2009; Joiner & Van Orden, 2008; Joiner, Van Orden, Witte, Selby, Ribeiro, Lewis, & Rudd, 2009; Joiner, Van Orden, Witte, & Rudd, 2009; Martin, Ghahramanlou-Holloway, Lou, & Tucciarone, 2009; Nademin, Jobes, Pflanz, Jacoby, Ghahramanlou-Holloway, Campise, Joiner, Wagner, & Johnson, 2008; Wingate et al., 2006). Despite the large number of applications of the theory by Joiner and his collaborators (e.g., Anestis et al., 2009; Joiner & Van Orden, 2008), core arguments in the original formulation of the IPT-ACS have not been closely examined. A review of the current status of the IPT-ACS reveals five issues that have not been addressed in post-2005 discussions of the theory.

First, in terms of Joiner's (2005) own narrative of some theories of suicide it appears that the IPT-ACS is not actually a new theory of suicide.

Second the suicidology literature citing Joiner's (2005) theory has not discussed omissions of studies in the original formulation of the theory (Joiner, 2005) that appear to contradict key elements in the IPT-ACS.

Third, another issue is the theory's attempt to explain the deaths of the 9/11 al-Qaida terrorists as suicides rather than martyrdom. In the same context, we also discuss the theory's monumental task to explain the deaths among certain victims of 9/11 attacks as a case of suicides rather than a case of homicides perpetrated by the al-Qaida terrorist organization.

Fourth, careful review of post-2005 qualitative and quantitative studies in support of the theory reveals that these studies have violated fundamental assumptions in Joiner's (2005) theory, suggesting that such studies have not actually empirically tested the theory.

Fifth, problems with the testability and falsifiability (Popper, 1935/1959) in the empirical evaluation of key variables in the IPT-ACS have not been addressed.

The above five issues have not only been ignored by Joiner and his collaborators in post-2005 discussions of the theory (e.g., Joiner et. al., 2009), but independent scholars have also ignored and/or avoided discussing such issues in the field of suicidology (Brenner, Gutierrez, Cornetter, Bethauser, Staves, 2008; Ellis, 2007; Simon & Hales, 2006).

The aim of this book is to evaluate the current status of the IPT-ACS with emphasis on the above five issues, hoping to stimulate further discussions regarding the merits and utility of the IPT-ACS in the field of suicidology. Before discussing each of these issues, however, it is important to provide a brief summary of three *key* psychological *constructs* in the IPT-ACS. It is also important to briefly discuss the role of suicidal ideation in the theory, which provides the background for understanding why the theory does not apply in the case of suicidal ideation in the absence of either attempted or completed suicide. As shown later, in this point the IPT-ACS has been misinterpreted in empirical studies testing the validity of theory. We conclude that rather than a theory, the IPT-ACS is a framework that could serve to unify the exemplars (Kuhn, 1996) of existing theories of suicide.

PSYCHOLOGICAL CONSTRUCTS
IN THE IPT-ACS

Joiner's theory (2005) postulates that a person who plans to attempt or commit suicide would have to experience *thwarted belongingness, perceived burdensomeness*, and the *acquired ability to enact lethal self-injury*. These are the three most important psychological constructs in the explanation of a suicide attempt or completed suicide. These three constructs must be present to explain the suicidal act (Anestis et al., 2009; Joiner, 2005; Joiner, & Van Orden, 2008; Martin et al., 2009). These constructs are briefly summarized below.

Thwarted Belongingness

According to the IPT-ACS, individuals would attempt suicide or kill themselves because they perceived they are not meaningfully connected to others, including family, friends, spouse, girlfriend or boyfriend, co-workers, and other valued groups. In this construct (*thwarted belongingness*), the individual perceives that prior established relationships have been lost.

Perceived Burdensomeness

Furthermore, individuals who attempt suicide or kill themselves do so because they also believe that they are a burden to their family, friends, loved ones, co-workers, community, society, and sometimes the entire world. In the IPT-ACS, when someone senses "the desire for death" (Joiner, 2005, p. 137) the individual is experiencing "two psychological states—perceived burdensomeness and failed belongingness" (p. 137).

It is important to point out that in the IPT-ACS the term "perceive" is critical, because Joiner (2005) wants to convey he is not talking about "actual burdensomeness" (p. 117). The theory makes the assumption that individuals who attempt suicide or kill themselves make a great mistake, because they perceive being a burden to other members of the community at the time they either attempt suicide or elect to die by suicide, when in fact that most likely is not the case.

Acquired Ability to Enact Lethal Self-injury

The third construct in the IPT-ACS postulates that, in addition to experiencing the two aforementioned two psychological states, individuals would finally attempt suicide or actually kill themselves if they had acquired *the ability to enact lethal self-injury* (see Joiner, 2005, pp. 46-93). How would they fulfill this psychological state? According to the IPT-ACS, they would have to experience *repeated* exposure to painful and provocative situations in the past leading to the ability to enact lethal self-injury in the future (see Joiner, 2005, pp. 56-59)

The key element in the third psychological state or construct is the *repetition* of aversive experiences that would provide the opportunity to "habituate to pain and provocation in general" (Joiner, 2005, p. 56). In the IPT-ACS, the distinction between "pain" and "provocation" is not clearly established. Examples of "pain" in the IPT-ACS would include history of childhood sexual and physical abuse, repeated tattooing and piercing, accidents, repeated self-injury with particular emphasis on multiple suicide attempts, and repeated exposure to war zone violence and atrocities (see Anestis et al., 2009, p. 46-93). An example of "provocative behaviors" (Joiner, 2005, p. 72) would be encounters with the legal system due to criminal behaviors (e.g., serious drug abuse, injury from recklessness, prostitution, and assault).

In the case of the third construct, the IPT-ACS predicts that "those who have habituated to pain and provocation through such means as serious drug abuse and prostitution should have demonstrably high suicide rates" (Joiner, 2005, p. 71). The reason for this observation is that the habituation to pain and provocation (through repeated exposures) results in "the diminution of fear" (Joiner, 2005, p. 58) to die by suicide. In Joiner's own words: "The acquired ability to enact lethal self-injury is a necessary precursor to serious suicidality, especially to completed suicide. This required ability involves fearlessness about confronting pain, injury, and indeed death" (Joiner, 2005, p. 92).

A unique feature of the third psychological state or construct is that, according to the IPT-ACS, one does not need to directly experience the repetition of pain and provocation in order engage in a future lethal self-injury (i.e., completed suicide). It is possible to become habituated to pain and provocative behaviors through the process of "vicarious habituation" (Joiner, 2005, p. 83; see also Joiner, Van Orden, Witte, & Rudd, 2009, p. 9).

Therefore, in addition to *directly experiencing* painful and provocative situations "people may habituate to dangerous stimuli…by observing someone else do so" (Joiner, 2005, p. 83). It should be noted that in the IPT-ACS "vicarious habituation" is not the same as "imitation" or "observational learning" in Bandura's social learning theory (e.g., Bandura, 1969). This is among the reasons for the absence of Bandura's theory of behavior in the IPT-ACS. The difference between "vicarious habituation" and the constructs of "imitation" or "observational learning" in Bandura's theory is further discussed in the testability or falsifiability section.

The Role of Suicidal Ideation in the ITP-ACS

Although the IPT-ACS recognizes the important role suicidal ideation plays in suicidal acts (see Joiner, 2005, p. 79), the IPT-ACS was formulated to explain suicidal acts (i.e., attempted suicide and completed suicide), and not suicidal ideation (see Joiner, 2005, p. 138, Fig. 1, and Joiner, Van Orden, Witte, & Rudd, 2009, pp. 4-14). The reason for the exclusion of suicidal ideation in the IPT-ACS is that people who only think about dying by suicide or talk about the desire to die (without attempt) do not engage in lethal self-injury (the third construct in the theory). If the individual engages in suicidal ideation and then postpones its materialization (i.e., does not attempt or commit suicide), and later engages in the suicide act, the IPT-ACS would be applied only in the case of the suicide act and not in the case of the suicidal ideation. As shown below, researchers have violated this exclusionary criterion (i.e., suicidal ideation) during a test of the IPT-ACS.

Summary

Joiner's (2005) theory is essentially "a reductionist approach in explaining why human beings [attempt suicide or] kill themselves" (Paniagua, 2010b, p. 297). The central argument in Joiner's theory is that before one engages in a lethal self-injury, one must first sense a "desire for death" (i.e., the psychological states of perceived burdensomeness and failed belongingness). As noted by Joiner, Van Orden, Witte, & Rudd (2009), the individuals must "experience the sustained co-occurrence" (p.

6) of both psychological states to actually sense a desire for death. When this desire for death "is combined with the acquired ability to enact lethal self-injury, the desire for death can lead to a serious suicide attempt or death by suicide" (Joiner, 2005, p. 136). In the discussion dealing with qualitative and quantitative research testing Joiner's (2005) theory, we show researchers have violated this fundamental assumption in the IPT-ACS (i.e., the demand to test the co-occurrence of the three psychological constructs in the same sample).

IS THE IPT-ACS A NEW
THEORY OF SUICIDE?

Joiner argued that the IPT-ACS is a new theory of suicide, relative to "numerous other theorists" (Joiner, 2005, p. 32) of suicide formulated during the past 113 years. Joiner (2005) also pointed out that he is "much a collaborator than a competitor with other theories" (p. 33) including, particularly, "the theories of Durkheim, Shneidman, Beck, Baumeister, and Linehan [that] are the most prominent and influential explanations of suicidal behavior" (Joiner, 2005, p. 42). Therefore, Joiner selected these five theories of suicide to justify the need for a new theory of suicide (i.e., the IPT-ACS). Below we show that the three constructs in the IPT-ACS have been previously considered in that selected sample of theories of suicide, but using different categorizations to name the same constructs. If true, the IPT-ACS may not be a new theory of suicide, but an approach that may serve to unify the exemplars (Kuhn, 1996) of existing theories of suicide.

Baumeister's Theory of Suicide.

Baumeister's theory (1990) postulates that the probability someone would die by suicide is dictated by the following steps: (1) the individual is confronted with "a negative and severe discrepancy between expectations and actual events" (Joiner, 2005, p. 40); (2) the individual "attempts *to escape from negative effect*, as well as from aversive self-awareness and the discrepancy between expectation and outcome" (Joiner, 2005, p. 40, italics added); (3) the individual accomplishes the second step "by retreating into a numb state of 'cognitive deconstruction'" (Joiner, 2005, p. 40), and (4) a significant "consequence of the state of cognitive deconstruction is reduced inhibitions, which contribute to lack of impulse control in general and lack of impulsive control for suicidal behavior in particular" (Joiner, 2005, pp.

40-41). In terms of the above steps, the individual would elect suicide as an alternative to escape from the actual events-expectations discrepancy. This is why this theory is also known as the "escape theory" of suicide (Baumeister, 1990, p. 98).

Joiner (2005) agreed the IPT-ACS is similar to Baumeister's theory when he wrote: "there are compatibilities between Baumeister's account and [his theory]. For example, perceived burdensomeness and failed belongingness can be seen as the result of disappointed expectations...that are internally attributed and thus associated with severe states of *negative affect*" (p. 41, italics added). In Baumeister 's (1990) escape theory of suicide, "certain self-attributions cause an awareness of self as inadequate, which leads to *negative affect*, especially depression and anxiety" (p.98, italics added). Escape theory, however, holds that whereas "there is clear evidence that depressed affect precedes suicide attempts [and completed suicide, in terms of overall logic of the theory]...the role of anxiety in the causation of suicide remains mainly *at the level of inference*" (Baumeister, 1990, p. 99, italics added). Therefore, one would assume that Joiner (2005) used the phrase "negative affect" to emphasize "depressed affect" in the escape theory of suicide (see Baumeister, 1990, p. 99), and not anxiety.

Furthermore, in Joiner's (2005) own interpretation of Baumeister's theory "cognitive deconstruction" is a critical construct in explaining why an individual elects suicide to escape from that discrepancy (see Baumeister, 1990, pp. 99-101). Joiner, however, pointed out that "the state of cognitive deconstruction is not part of the current model [i.e., IPT-CAS]" (Joiner, 2005, p. 41). Despite this exclusionary conclusion, it appears that Joiner realized that the compatibility between his theory and Baumeister's theory could not be justified with the exclusion of that construct in Joiner's theory. Therefore, Joiner (2005) continued, "...but one could imagine that perceived burdensomeness and failed belongingness are painful enough to produce such a condition [i.e., cognitive deconstruction]" (p. 41). This conclusion allowed Joiner to show the IPT-ACS and Baumeister's theory are inter-connected, in that cognitive deconstruction results from thwarted burdensomeness and belongingness.

Beck's Theory of Suicide

In the case of Beck's theory, the construct of "hopelessness" is crucial in explaining why people die by suicide (Beck, Brown, & Steer, 1989; Beck, Brown, Berchick, Stewart, & Steer, 1990; Brown, Jeglic, Henriques, & Beck, 2006). Joiner reviewed empirical findings published by Beck and his collaborators showing people with high hopelessness (as measured by the Beck Hopelessness Scale) are "more likely to die by suicide as compared to [individuals] with lower hopelessness scores" (Joiner, 2005, p. 39). Joiner (2005), however, argued that hopelessness in *itself* cannot explain suicide acts (attempt and completed suicide) and that this construct must be combined with the three constructs in the IPT-ACS to explain suicide acts: "*hopelessness* about belongingness and burdensomeness is required, together with the acquired capability for serious self-harm" (p. 39, italics added). In this argument, however, Beck's hopelessness construct is still a key factor.

Joiner (2005) also agrees his mechanism of "habituation, or getting used to the fear and pain involved in self-injury" that eventually "leads to an acquired ability for self-injurious suicidality" (p. 40), is similar to the mechanism of "cognitive sensitization" (p. 40) in Beck's theory. Joiner pointed out that in Beck's theory, this mechanism emerges when the *repetition* of "previous suicidal experience sensitizes suicide-related thoughts and behaviors such that they later become more accessible and active. The more accessible and active the thoughts and behaviors become, the more easily they are triggered, and the more severe are the subsequent suicide episodes" (Joiner, 2005, p. 39).

As noted above, in the IPT-ACS the repetition of aversive experience is a key factor leading to the "habituation" to "pain and provocation" that eventually results in the "acquired ability to engage in serious self-injury" (Joiner, 2005, p. 47). Similarly, in the case of Beck's theory Joiner (2005) pointed out that the repetition of "suicide-related thoughts and behaviors" (p. 40) is also a key factor for the presence of the "cognitive desensitization" construct. Joiner (2005) concluded: "these mechanisms *are not mutually exclusive* and thus may operate jointly" (p. 40, italics added). Therefore, Joiner's conclusion is that both theories deal with the same construct, but labeled differently, namely, "habituation" (Joiner's theory) and "cognitive desensitization" (Beck's theory).

Brown et al. (2006) reported that in order to accommodate research development in theoretical analyses of suicidal behavior, Beck (1996; cited by Joiner, 2005, p. 39) revised his theory of suicide with the inclusion of a new construct, namely, *modes*. As noted by Brown et al. (2006), "Beck defines modes as interconnected networks of cognitive , affective, motivational, physiological, and behavioral schemas that are activated simultaneously by relevant internal and external events and orient the individual toward achieving some goals" (p. 60). Brown et al. (2006) also pointed out that "Beck introduced the concept [or construct] of modes to account for the diversity of symptoms observed in most psychiatric conditions" (p. 60).

Furthermore, Brown et al. (2006) pointed out that although Beck "touched on a suicide mode" (p. 61), its role in explaining suicide was further elaborated by Rudd (2000) and Rudd, Joiner, and Rajab (2001). These authors "postulated that when the suicide mode is activated, an individual experiences suicide-related cognition, negative affect, physiological arousal, and the motivation or intent to engage in suicidal behavior" (Brown et al., 2006, p. 61). Because Joiner was involved in further elaboration of the construct of "modes" (see Rudd et al., 2001, pp. 24-33), one would wonder why Joiner did not use Beck's (1996) revised theory of suicide in which Beck postulated modes "to account for the diversity of symptoms observed in most psychiatric conditions" (Brown et al., 2006, p. 60), particularly involving suicide. Perhaps Joiner felt Beck's modes were not compatible with the three psychological states in the IPT-ACS.

Durkheim's Theory of Suicide

A core construct in Durkheim's (1897/1951) theory of suicide is the role of social integration as an explanation of why people die by suicide. As noted by Joiner, in Durkheim's theory "low integration [to society] ... leads to an increase in a type of suicide that Durkheim labeled 'egoistic'" (Joiner, 2005, pp. 33-34; see also Durkheim, 1897/1951, pp. 152-216; Lester, 2004b; Lester, 2008). The construct of low integration is "referred to as low belongingness" (Joiner, 2005, p. 33) in the IPT-ACS. On the other hand, "too much integration [to society], according to Durkheim, is also associated with more suicide" (Joiner, 2005, p. 34). But in this case, a different type of suicide was postulated by Durkheim: *altruistic suicide* (see Durkheim, 1987/1951, pp. 217-240; Lester, 2008, pp. 45-46).

Joiner (2005) pointed out that a crucial element of altruistic suicide is "self-sacrifice" and concluded that "self-sacrifice bears some similarities to the concept of perceived burdensomeness" (p. 34). As noted by Joiner, "through his [i.e., Durkheim] emphasis on altruistic suicide, he also anticipated my theory's inclusion of perceived burdensomeness as a key precursor to serious suicidal behavior" (p. 35). Therefore, Joiner (2005) has labeled the two processes postulated by Durkheim with two different categorizations: (1) thwarted belongingness (the same as "low integration" to society leading to "egoistic suicide"), and (2) perceived burdensomeness (the same as "self-sacrifice" in individuals experiencing "altruistic suicide" because of "too much integration" to society).

Another significant similarity between Durkheim's and Joiner's theory is that Durkheim "did not deny …that individual conditions *like mental disorders* are relevant to suicide. But he did claim that most such [individual conditions] are insufficiently general to affect the suicide rate of whole societies, and thus should not be emphasized by *sociologists*" (Joiner, 2005 pp. 34-35, italics added). Joiner's theory concluded exactly in the same way, except that the IPT-ACS encourages *psychologists* not to emphasize the role of mental disorders in the explanation of suicidal acts. This observation is further discussed below.

Linehan's Theory of Suicide

Joiner (2005) pointed out that "the current framework [Joiner's theory] and Linehan's model [993] are …quite compatible" (Joiner, 2005, p. 42; see also Brown, 2006). Here is how Joiner arrived at this conclusion: "Marsha Linehan has theorized that biological deficits, exposure to trauma, and the failure to acquire adaptive ways of tolerating and handling negative emotion all contribute to suicidal behavior" (Joiner, 2005, p. 41). According to Linehan's theory of suicide, to deal with these situations the individual would first engage in "emotional regulation mechanisms" (Joiner, 2005, p. 41) and when these mechanisms fail, the next step is to engage in "self-injury…in an attempt to regulate emotions" (Joiner, 2005, p. 41). When these mechanisms "have broken down or never developed adequately" (Joiner, 2005, p. 41), the resulting negative outcome is "emotional dysregulation" (p. 42; see also Linehan, 1993, pp. 42-65) that leads to "the acquired capability to enact lethal self-injury" (Joiner, 2005, p. 42). Therefore, in the theoretical explanation of why some individuals engage in lethal self-injury the IPT-ACS is similar to Linehan's (1993) model.

It is also important to note that, in addition to the construct of "emotional dysregulation," Linehan also postulated that this construct must interrelate with the construct of "self-invalidation" that "*together* lead to suicide behavior" (Brown, 2006, p. 98, italics added). Subjective experiences such as low self-esteem, self-directed anger, negative self-judgments, and shame are examples of self-invalidation in Linehan's theory. As noted by Brown (2006), "suicidal individuals invalidate themselves when they blame and judge themselves harshly for the lack of control of behavior and emotions, and when they treat their normal responses as invalid" (p. 98). This self-invalidation construct, however, does not play a role in the IPT-ACS. Perhaps Joiner (2005) realized he could not provide a convincing rationale for equating the construct of "self-invalidation" with any of the three constructs in the IPT-ACS, as he did with the construct of "emotional dysregulation."

It is important to point out that Linehan used Bandura's social learning theory (e.g., Bandura, 1977) to explain how "self-invalidation develops when suicidal individuals learn to disregard, criticize, and punish themselves (and their emotions) from *observing others* who have invalidated them in these ways" (Brown, 2006, p. 101; italics added). Joiner (2005), however, did not cite or discuss Bandura's theory, suggesting that Bandura's social learning theory did not have a role in the formulation of the IPT-ACS. By electing to avoid the construct of "self-invalidation" in his theory, however, Joiner (2005) actually did not explain why people die by suicide according to Linehan's theory of suicide (see Brown, 2006, pp. 92-100). The reason for this conclusion is that, as noted by Brown (2006), in Linehan's theory "emotional dysregulation and self-invalidation are *interrelated* problems that *together* lead to suicidal behavior" (p. 98, italics added).

Shneidman's Theory of Suicide

Shneidman (1996) argues that "in almost every case, suicide is caused by pain, a certain kind of pain---psychological pain, which I call *psychache (sik-ak)*. Furthermore, this psychache stems from thwarted or distorted psychological *needs*" (p. 4, italics in the text). Joiner (2005) then wrote: "an emphasis on psychache...is compatible with my approach; perceived burdensomeness combined with failed belongingness constitutes psychache" (p. 37). Joiner's conclusion in this context is confirmed with the three clinical cases (i.e., Ariel Wilson, Beatrice Bessen, and Castro Reyes) Shneidman (1996) used to support his theory of suicide. Shneidman titled the third case, "The need to belong: The case of Castro Reyes," which deals with the construct of "failed belongingness" (see Joiner, 2005, pp. 117-134).

Shneidman (2004) also described the case of Arthur, a 33-year-old white male physician-lawyer who killed himself following a long history of severe depression and suicide attempts. In Shneidman's narrative of Arthur's case, the theoretical constructs of "failed belongingness" (e.g., Shneidman, 2004, p. 94) and "perceived burdensomeness" (e.g., Shneidman, 2004, p. 139) are evident across the entire case. Actually, in the discussion of Arthur's case Shneidman pointed out that among the "four central themes to the suicide belief system" (Shneidman, 2004, p. 139), namely, *"unlovabilty* ("I don't deserve to live"); *helplessness* ("I can't solve my problem"); *poor distress tolerance,* or psychache ("I can't stand this pain any more); and *perceived burdensomeness* ("Everyone would be *better off* if I were dead")" (Shneidman (2004, p. 139, italics added), Arthur's case clearly met the fourth central theme: *"perceived burdensomeness"* (Shneidman, 2004, p. 139. italics added). Joiner (2005) agreed with Shneidman when he wrote: "Making others *better off* is similar on concept to perceived burdensomeness" (p. 108, italics added). Therefore, in terms of Joiner's own interpretation of Shneidman's (1996) theory of suicide, it is evident that Shneidman (1996, 2004) preceded Joiner (2005) in the formulation of the constructs of failed belongingness and perceived burdensomeness.

Joiner (2005) further pointed out that Shneidman's theory of suicide also "identified lethality as a key ingredient of serious suicidality. Lethality [as proposed by Shneidman's theory] is clearly related to the....acquired ability to enact lethal self-injury" (p. 37) in the IPT-ACS. As noted earlier, in the IPT-ACS the process of "habituation to pain and provocation" (Joiner, 2005, p. 56) is required for the individual's "acquired ability to enact lethal self-injury" (Joiner, 2005, p. 57). The case of Ariel Wilson described by Shneidman (1996) on pages 30-44 shows that Wilson experienced lots of "pain and provocation" before she "attempted to take her life by setting herself on fire" (Shneidman, 1996, p. 27). The same observation can be applied in the cases of Beatrice Bessen (pp. 67-80) and Castro Reyes (pp. 97-125) in Shneidman's (1996) theory of suicide. This construct is also evident in Arthur's case, particularly in several suicide attempts prior to his death by suicide (e.g., see Shneidman, 2004, p. 167).

The only reservation Joiner (2005) reported in the case of Shneidman's (1996) theory of suicide is that the answer to the question, "what in *particular* ...are people feeling psychache about?" (Joiner, 2005, p. 38, italics added) "is too general" (p. 37) because Shneidman answered that question by adapting into his theory all the thwarted needs postulated by "[Henry A.] Murrays' work in the 1930s" (Joiner, 2005, p. 37). On page 37, Joiner (2005) listed 20 thwarted needs Shneidman (1996, see p. 20) selected from Murray's (1938) work as potential explanations for suicide acts. Joiner (2005) argued that all 20 psychological needs are unnecessary and that a better answer to that question would be that people feel psychache when they have "perceived burdensomeness and failed belongingness" (Joiner, 2005, p. 38).

Shneidman's (1996) theory of suicide, however, does not demand *all 20* thwarted needs postulated by Murray (1938) are needed to explain "why do we kill ourselves" (see Shneidman, 1996, pp. 3-26, Chapter 1). For example, in the case of Ariel Wilson, Shneidman (1996) wrote: "We can use a method that helps us place values on the 20 psychological needs and note how they shape Ariel's [Wilson] personality. What is important is seeing which needs *predominated*" (p. 46, italics added). Shneidman (1996, see p. 46) concluded that, among the 20 psychological needs cited by Joiner (2005) on page 37, only three were predominant in explaining Ariel Wilson's suicide attempt: *succorance, deference, and nurturance.*

The first construct (succorance) means "to have one's needs gratified; to be loved" (Shneidman, 1996, p. 20); the second (deference) is "to vindicate the self against criticism or blame" (p. 20), and the third construct means to "defeat, help, console, protect, nurture and ideas" (p. 20). Additional definitions of 17 additional needs postulated by Murray (1938) can be found in Shneidman (1996, see p. 20). Furthermore, Shneidman (1996) pointed out that, among these psychological needs, "Ariel's psychological pain was driven *primarily* by her need for succorance, to be loved" (p. 68, italics added). This *need* for succorance corresponds to Joiner's construct of "failed belongingness in suicide" (e.g., Joiner, 2005, p. 120). Therefore, the constructs of succorance need and failed belongingness are essentially the *same in both theories.*

It should be noted that another theory of suicide proposed by Shneidman is known as "The Cubic Model of Suicide" (see Shneidman, 1987, pp. 174-177). Jones and Nelson (2006) summarized this "cubic model" of suicide in the following terms: "The first axis in [the] cubic model is unbearable psychological pain [the construct of psychache described above] that can be rated from *low* (1) to *high* (5). The second axis is that of unrelenting psychological pressures...or stressors that can be rated from low to high... The third axis is...perturbation [e.g., emotionally disturbed and upset], also rated from low to high" (p. 35, italics in the text). Shneidman (1987) postulated that an individual would commit suicide when he/she is at the highest levels of perturbation, pain, and stress or "at the 5-5-5 corner cubelet of the model" (Jones & Nelson, 2006, p. 35). This "cubic model" of suicide does not appear to have enough theoretical ingredients ---with the exception of the psychache construct---to find commonalities between the IPT-ACS and Shneidman's cubic model. This might explain why Joiner (2005) did not mention this "cubic model" during the formulation of the IPT-ACS.

Summary

Joiner only selected five out of over fifteen theories of suicide (Lester, 1990; Lester, 2004a; Lester, 2004b) to support the formulation of the IPT-ACS. These five carefully selected theories provided the concepts and process needed to support the rationale for the selection of the three constructs in the IPT-ACS. Other theories of suicide Joiner (2005) did not consider provide less support for the three constructs in the IPT-ACS including, for example, the anthropological (Lester, 2004b; Lester, 2008), cross-cultural (Leenaars, 2008), ethological (Lester, 2004b; Goldney, 2001), social conflict (Lester, 1990), and stress-diathesis (Heeringen, 2001b; Mann, Waternaux, Haas, & Malone, 1999) theories of suicide.

OMISSIONS IN THE LITERATURE
CITED IN THE 2005 IPT-ACS

Joiner (2005) cited the literature extensively to support the claim that his three key constructs in the IPT-ACS is the best explanation of suicide attempts and, particularly, completed suicide. Only some examples from that literature are discussed below, to illustrate the fact that in the original formulation of the IPT-ACS Joiner ignored competing arguments regarding why people either attempt or die by suicide.

Suicide Rates Among Twins

Tomassini, Juel, Holms, Skytthe and Christensen (2003) reviewed empirical findings suggesting that the rate of suicide is generally lower among twins, relative to non-twins. Joiner (2005) cited Tomassini et al. (2003) and then explained these findings by arguing that "if belongingness is implicated in suicidality, one might predict that twins enjoy some protection from suicide, given the belongingness inherent in twinship" (p. 126). By the logic of the IPT-ACS, one would also assume that twins in the study Joiner cited (i.e., Tomassini et al. 2003) *were not* "habituated" to "pain and provocation" leading to the "acquired capability to engage in serious self-injury" (Joiner, 2005, p. 47), and also did not report they were a burden to their parents, friends, and other individuals in their life. Joiner (2005), however, reported a fact, namely, lower suicide rate among twins, relative to non-twins, but he did not provide empirical evidence in support of the presence of the three constructs in the IPTACS associated with that fact. In addition, evidence for these constructs is also missed in the study by Tomassini et al. (2003).

Furthermore, during the report of the above fact, Joiner (2005, see p. 126) did not review findings from other twin studies evidencing a genetic effect associated with death by suicide among twins (Jamison, 1999; Roy, Nielsen, Rylander, Sarchiapone, & Segal, 1999; Roy, Rylander, & Sarchiapone, 1997). For example, Roy et al. (1997) reviewed a series of twin studies reporting higher concordance rates for suicide in identical twins (monozygotic twins) than in non-identical ones (dizygotic twins), strongly suggesting genetic influences in the explanation of these findings (similar results can be found in Roy et al., 1999, p. 13). The results reported by Roy et al. (1999) suggest the *inapplicability* of the IPT-ACS in this context, because a substantial number of identical twins committed suicide despite the assumption in the IPT-ACS regarding that one would predict a lower concordance rate for suicide among identical twins due to the "fact" that they "enjoy some protection from suicide, given the belongingness inherent in twinship" (p. 126).

Therefore, Joiner simply cited Tomassini et al. (2003) to confirm the IPT-ACS in this context, but he failed to cite contradictory evidences against the IPT-ACS (e.g., Roy et al., 1997; see also Goldney, 2001; Heeringen, 2001a; Heeringen, 2001b; Heeringen, 2001c; Williams & Pollock, 2001). In this discussion, it is important to remember the following passage from Williams and Pollock (2001): "...it is not enough for psychologists [e.g., Joiner] to build their explanations [of suicidal behavior] *in isolation*. It is important to keep in mind how psychological explanations fit with social facts...and biological/genetic facts...We are not engaged in a race in which either social, psychological or biological [/genetic] facts will win. Instead, we must look for how our different facts and theories can fit together" (p. 77, italics added).

Suicide Rates Among Immigrants

The IPT-ACS predicts that the suicide rate among immigrants should increase because of "separation from a 'mother country,' according to a *belongingness* view, might be associated with heightened suicidality" (Joiner, 2005, p. 127, italics added). Joiner based this conclusion on a study by Yampey (1967) that reported a "very high rate of suicide... among foreign-born males" (p. 127) in the late 1800s in Buenos Aires, Argentina. In that conclusion, however, Joiner did not report findings regarding the presence of the other two constructs in the theory. In terms of Joiner's (2005) theory, one would speculate that because those "foreign-born males" died by suicide, one would have to assume (in the absence of empirical evidence, of course) that they also had the *"capability to engage in serious self-injury"* (Joiner, 2005, p. 47, italics added) and perceived they were *a burden* to their family, friends, co-workers and other individuals they left behind in their "mother country" (Joiner, 2005, p. 127). The reason for this conclusion is that, as noted earlier, these three constructs (in italics above) *must* be present to explain deaths by suicide (Anestis et al., 2009).

In addition, Joiner (2005) did not reveal two important points regarding the Buenos Aires study he cited on page 127 (i.e., Yampey, 1967). First, this study is characterized by several methodological deficiencies (e.g., sampling procedures and reliability measures were not reported). Second, Yampey (1967, see p. 43) considered the role of medical complications, family conflicts, economic difficulties, loss of loved ones in the family, marital problems, and mental disorders in the explanation of findings, especially "in people who came [to Buenos Aires, Argentina] from Europe" (Yampey, 1967, p. 44). Therefore, in this Buenos Aires case Joiner (2005) failed to report significant factual details in Yampey's article, particularly Yampey's own explanation of results beyond "the intense social mobility" (Yampey, 1967, p. 44) in Buenos Aires from 1895 through 1965 (the time period covered by Yampey in his study; see Yampey, 1967, Table 4, p. 42).

Suicide Rates in Times of National Tragedies

Joiner (2005) reviewed the literature suggesting that the rate of suicide tends to decrease during national tragedies (see Joiner, 2005, pp.127-129). According to the IPT-ACS, the reason for this fact is that "in times of acute national crisis, people pull together, and belongingness should increase" (p. 127). In terms of the logic of the IPT-ACS, in these cases the ability to enact lethal self-injury and perceived burdensomeness should thus decrease, otherwise the IPT-ACS would not be able to explain that fact. Joiner (2005), however, reported a fact documented in a particular study (cited in Joiner's book, 2005), but he did not provide empirical findings to show the role of the three constructs in explaining the suppression of suicidal behavior in times of crisis.

An example of findings against the explanatory power of the IPT-ACS in cases of national tragedies can be found in Chou, Huang, Lee, Tsai, Tsay, Chen, and Chou (2003), but unreported by Joiner (2005). In brief, Chou et al. (2003) found that after the 1999 Taiwan earthquake victims of this national tragedy were more likely than non-victims to commit suicide. This finding suggests these victims did not "pull together" and that "belongingness" decreased among them, which is the opposite assumption in the IPT-ACS.

Suicide Rates Among Mentally Ill Individuals

With the exception of Durkheim's theory of suicide (1897/1951), the other "prominent and influential explanations of suicidal behavior" (Joiner, 2005, p. 42) Joiner cited as a theoretical background to build the IPT-ACS (e.g., Beck, 1976; Shneidman, 1996) agree that mental disorders play a key role in the rate of suicide attempts and completed suicide (see also Ellis, 2007; Heeringen, 2001a; Moscicki, 2001; Shneidman, 2004; Simon & Hales, 2006). Joiner (2005) agreed with this observation when he wrote that "approximately 95 percent of people who die by suicide experienced a mental disorder at the time of death" (p. 191). A similar conclusion can be found in Jamison (1999, p. 100) and Moscicki (2001, pp. 315-316). Joiner (2005), however, concluded that the three constructs of the IPT-ACS provide the best explanation for increasing rates of suicide among mentally ill individuals, and not mental disorders per se.

For example, Joiner (2005) agreed that mood disorders (e.g., Major Depressive Disorder and Bipolar Disorders) "deserve consideration in any discussion of suicide" (p. 199). But what explains the "high suicide rates in mood disorders may be a function of the ability to enact lethal self-injury" (p. 199) in combination with "perceived burdensomeness, and failed belongingness" (p. 199). Similarly, the rates of suicide attempts or completed suicide among individuals who engage in Substance-Related Disorders are associated with the fact that these disorders can facilitate "provocative experiences and thus the acquisition of the ability to enact lethal self-injury" (p. 193.

In addition, "prolonged substance abuse can certainly deteriorate social capital (leading to low *belongingness*) and diminish feelings of overall effectiveness (producing feelings of perceived *burdensomeness*)" (p. 194, italics added), which could also impact on the increasing rates of suicide attempts and completed suicide among individual experiencing substance abuse disorders. Additional examples of mental disorders the IPT-ACS discusses in the present context include Anorexia Nervosa (Joiner, 2005p. 72, p. 195), Borderline Personality Disorder (p. 72, p. 195), and Panic Disorder with Agoraphobia (p. 193).

The overall conclusion in the context of explaining rates of suicide attempts or completed suicide among mentally ill individuals is that it is true that "virtually everyone who dies by suicide [about 95%] experienced one or more mental disorders at the time of their death" (Joiner, 2005, p. 201), but the IPT-ACS argues that the rates of suicide attempts and completed suicide among mentally ill individuals are best explained with this theory and not by psychopathological explanations.

During the formulation of the IPT-ACS, however, Joiner (2005) did not cite the literature with emphasis on findings from the *Standardized Mortality Ratio* published prior to 2005. In addition, in post-2005 publications with emphasis on the IPT-ACS (e.g., Joiner et al., 2009; Joiner, Van Orden, Witte, & Rudd, 2009) Joiner and his collaborators have not considered findings from *Population Attributable Risk* studies that would compete with the IPT-ACS in the context of explaining the same phenomenon (i.e., rates of suicidal behavior among mentally ill individuals).

Harris and Barraclough (1997) used the Standardized Mortality Ratio[1] (SMR, given in parentheses below) to determine which mental disorders might best predict suicidal behavior and found the following disorders had the highest SMR: Eating Disorder (23.14), Major Depressive Disorder (20.35), Sedative Abuse (20.34), Mixed Drug Abuse (19.23), Bipolar Disorder (15.05), and Opioid Abuse (14.00). An extensive discussion of these findings with additional SMRs across other mental disorders can be found in Simon (2006, see p. 18).

Similarly, findings have been reported using the Population Attributable Risk (PAR)[2] to estimate the impact of a wide range of mental disorders on suicidality. For example, Krysinska and Martin (2009) reviewed the literature on PAR associated with suicidal behavior and found that "affective, anxiety, and substance use disorders are important contributing factors for suicidality, with PAR fractions reaching values of 40% to 80% (affective disorders), 7% to 50% (anxiety disorders), 8% to 60% (substance use disorders), and 23% to 46% for comorbidity between mental disorders and substance abuse" (p. 534). These findings from the PAR have not been considered in post-2005 discussions involving the IPT-ACS (e.g., Joiner et al., 2009). Regardless of the particular theory of suicide under consideration (Lester, 2004b; Simon, 2006), including Joiner's (2005) theory, these findings are very difficult to deny in the context of predicting who is at high risk for suicidal behavior.

Sadly, in terms of the overall assumptions in Joiner (2005) theory, we decided that it was important to point out that a crucial observation that might be against Joiner's diminished role of mental disorders in the present context is Joiner's explanation of the death of his father by suicide. Joiner agreed that psychopathology was a key risk factor in the explanation of why his father killed himself. In addition, in the Prologue Joiner (2005) suggested that his theory cannot explain his father's death by suicide.

If true, his father's death by suicide is the only case in the entire world that could not be explained with the IPT-ACS. The reason for this sad conclusion is that, as pointed out earlier, Joiner himself wrote that his theory is the only one that can explain "all deaths by suicide worldwide, across cultures" (Joiner, 2005, p. 226).

In the Prologue of his book, Joiner (2005) revealed that Bipolar II Disorder "...a serious depression combined with hypomanic episodes... played a role in his [father's] death" (p. 191). In addition, it appears that only two constructs in Joiner's theory were directly associated with his father's death by suicide. For example, in the Prologue Joiner provided ample evidences to show that his father "acquired ability to enact lethal self-injury "(see Joiner, 2005, pp. 12-13). On page 160, Joiner also discussed a scenario suggesting that his father also "failed belongingness." The construct of "perceived burdensomeness" is not apparent in the Prologue, suggesting that Joiner's father did not perceive to be a burden to his family. Actually, Joiner wrote: "my own dad died by suicide and the idea that he was an actual burden *is offensive*" (p. 115, italics added). As noted earlier, the theory has explanatory power only in those cases where the suicidal person engaged in each of the three constructs (see Joiner, 2005, p. 202).

Furthermore, in the Prologue Joiner (2005) appears to suggest that his theory plays a secondary role in explaining the death of his father, because Joiner emphasized a long history of untreated depression as the most parsimonious explanation for the death of his father by suicide. Examples in support this observation include: "dad had a mood disorder, and one of relatively long duration" (Joiner, 2005, p. 9); when he died "he was in the middle of a depressive episode" (p. 9); "near the end of his life, he seemed to have even more obvious depressive episodes" (p. 9); "approximately as many people with bipolar II die by suicide as those with bipolar I...and I now see that this was what my dad had" (p. 11); "during all these years my dad had these symptoms" (p. 11). Therefore, it appears that in the case of Joiner's father mental disorders were the primary factors for his death by suicide and not the tenet of Joiner's theory of suicide.

Suicide Notes and the IPT-ACS

Although not considered empirically tested procedures but generally anecdotal reports (Shneidman, 2004), suicide notes might also be more convincing in providing insights regarding the reasons for suicidal behavior, relative to the three constructs in the IPT-ACS This observation was not thoroughly discussed by Joiner (2005).

All theorists of suicide typically would agree that suicide notes left by individuals who kill themselves are the best explanation for their death by suicide (Shneidman, 2004). The suicide note the American artist Ralph Barton (1891-1931) wrote before he killed himself on May 19, 1931 is one of many examples of suicides notes illustrating the role of psychopathology among people who die by suicide. Barton wrote: "Everyone who has known me and who hears of this will have a different hypothesis [theory] to explain why I did it. Practically all of these hypotheses [theories] will be dramatic—*and completely wrong.* Any sane doctor knows that the reasons for suicide are invariably psychopathological" (Jamison, 1999, p. 85, italics added). In Barton suicide's note, he reported that "definitive symptoms of manic-depressive insanity" (Jamison, 1999, p. 86) was the main reason for his death by suicide. A similar case was published by Shneidman (2004), in which a white male physician-lawyer ("Arthur") killed himself after a long history of severe depression. In his suicide note, Arthur wrote: "Depression has slowly eaten away at my life" (Shneidman, 2004, p. 165), adding that "my illness is a tragedy, but it is one that I unfortunately cannot overcome" (p. 170), except by killing himself.

Barton's suicide note includes words suggesting that he did not experience the three constructs in the IPT-ACS. Barton wrote (Jamison, 1999, pp. 85-86): "I have had few difficulties...I have always had excellent health" [i.e., absence of habituation *to pain and provocation*]. I have had, on the contrary, an exceptionally glamorous life...And I have had more than my share affection and appreciation [i.e., absence of *perceived burdensomeness*]. The most charming, intelligent, and important people I have known have liked me" [i.e., absence of *thwarted belongingness*]. Barton concluded that the main reason for killing himself was the "definite symptoms of manic-depressive insanity" (Jamison, 1999, p. 86) he could not handle anymore.

Summary

In the preceding discussion, it appears that Joiner (2005) used a selective approach in terms of making sure that examples from the suicidology literature would confirm the theory. We showed that the domains Joiner selected (i.e., suicide rates among twins and immigrants, during national tragedies, and among mentally ill individuals) to review the existing literature at the time he wrote the theory did not include findings that appear to contradict the central core of Joiner's (2005) theory of suicide.

A THEORY OF EVERYTHING: SUICIDES VERSUS HOMICIDES DURING THE 9/11 ATTACKS

Joiner (2005) went to great lengths in trying to demonstrate how the IPT-ACS explains the consequences of the 9/11 attacks. For this reason, the IPT-ACS is obviously a macro-level theory (Creswell, 2009) that could now be re-named the "theory of everything," because if the IPT-ACS can explain the death of all 9/11 al-Qaida terrorists and certain victims of 9/11) as suicides this theory would not have difficulty explaining suicidal behavior in all imagined cases in the entire world. Actually, as noted earlier, Joiner (2005) arrived at a similar conclusion when he wrote that the IPT-ACS has "something to say about all deaths by suicide *worldwide,* across all cultures, by employing three simple concepts [constructs]" (p. 226, italics added).

Did the 9/11 al-Qaida Terrorists Die by Suicide?

In the case of the 9/11 al-Qaida terrorist attacks, Joiner (2005) asked, "Did they die by suicide?" (p. 142). This is not the right question to ask, according to experts on terrorism (e.g., Ditzler, 2004; Habeck, 2006; Marsella, 2004; Moghaddam, 2006; Moghaddam & Marsella, 2004). Members of the al-Qaida terrorist organization "do not conceptualize 'suicide bombers' as committing suicide-'martyrdom' is a preferred term" (Lester, 2008, p. 40). The reason for an act of martyrdom rather than suicide in the mind of 9/11 al-Qaida terrorists is that they believed their terrorist acts will help al-Qaida to achieve two fundamental goals, namely, to kill disbelievers of the revelations of the Prophet Mohammad in the Glorious Qur'an, and to establish an Islamic state governing the entire world (Emerson, 2006; Habeck, 2006; Moghaddam & Marsella, 2004;

Paniagua, 2005; Paniagua, 2006; Paniagua, 2007).[3] Actually, Joiner (2005) agreed with this interpretation of "martyrdom" when he wrote that during the 9/11 attacks "…terrorists themselves would more likely have characterized their deaths as martyrdom or casualties of holy war than as suicide" (p. 28).

In the same narrative of al-Qaida terrorist attacks on 9/11, however, Joiner (2005) changed his mind regarding his explanation for the motivation of the attacks as "martyrdom or casualties of holy war than as suicide, "and then argued that the 9/11 al-Qaida terrorists died by suicide and not by martyrdom. Joiner (2005 asked, "How does the current model [IPT-ACS] understand them?" (p. 142). According to the IPT-ACS, the terrorists experienced pain and provocation during their "training and preparation for the [terrorist] act" (Joiner, 2005, p. 142). This training led to the acquired "capability for lethal behavior" (p. 142).

In addition, the 9/11 al-Qaida terrorists also believed (before the attacks) that "their death is worth more to their community than their life" (Joiner, 2005, p. 142), which is a "kind of calculation related to burdensomeness" (p. 142). Joiner (2005), however, encountered an obstacle in the use of the construct "thwarted belongingness" to finally explain the 9/11 terrorists' attacks as suicides. Joiner (2005) first recognized the 9/11 terrorists were attached to their organization, and for this reason "their sense of belongingness…seemed quite high" (Joiner, 2005, p. 142). This is a fact reported in the al-Qaida terrorist literature (e.g., Habeck, 2006; Moghaddam & Marsella, 2004; Paniagua, 2005). Joiner's conclusion, however, does not support the applicability of one of the three main constructs in the theory, namely, that a sense of belongingness should be low for the theory to explain why some individual engage in suicidal acts. Joiner (2005), however, found a way to show that the construct of "thwarted belongingness" also applies in the present situation. Joiner (2005) wrote: "It is possible that death and life merge for suicide terrorists, such that they view death as a way to belong and to belong more fully than by anything they could do in life" (p. 142). In this discussion, we conclude that Joiner's (2005) interpretation of the 9/11 al-Qaida terrorist attacks as an example of suicidal behavior is speculative and empirically baseless, for the following three reasons.

First, Joiner (2005) did not provide evidence (e.g., through psychological autopsies conducted with relatives and friends of the 9/11 terrorists) to assess the three constructs in the IPT-ACS. The theory simply asserts that if the 9/11 attacks were "suicidal" attacks, the three constructs in the theory will be needed to explain those attacks as "suicides," but in the absence of empirical evidence to support this conclusion.

Second, it is true that all members of the al-Qaida terrorist organization are habituated to pain and provocation during their training (Moghaddam, 2006), but it is equally true that a very small number of those trainees who engaged in habituation to pain and provocation participated in the al-Qaida terrorist attacks. For example, Stern (2008) reported that "by September 11, 2001, between 70,000 and 110,000 radical Muslims had graduated from al-Qaeda training camps" (p. 164), but only 19 executed the 9/11 attacks.

Third, the literature on the al-Qaida terrorist organization strongly suggests that individuals who join this organization do so *not because they sense* "failed belongingness" and "perceived burdensomeness" among relatives and friends, but because of their desire to help al-Qaida to achieve the above two aforementioned goals: kill disbelievers of the Prophet Mohammad's revelations *and* to have an Islamic state in the entire world (Emerson, 2006; Habeck, 2006; Marsella, 2004; Moghaddam, 2006; Moghaddam & Marsella, 2004).

It should be noted that the majority of terrorist attacks by al-Qaida are aimed at killing others and *staying alive* to plan and execute further attacks; martyrdom is reserved for a very small number of trainees (Stern, 2008), and when they die during the attacks, the reason for their death is not based on the three constructs of the IPT-ACS but, as noted by Mahan and Griset (2008, p. 142), on their belief that they "are donating themselves for the cause [of the organization.]"

Did Certain Victims of 9/11 Die by Suicide?

Joiner (2005) used his theory of everything to explain why certain victims during the 9/11 attacks jumped to their death from the upper floors of the World Trade Center (WTC). According to Joiner's (2005) theory, those deaths can be explained by the three constructs in the theory, rather than by homicides resulting from the 9/11 al-Qaida terrorists attacks. This explanation would be perceived as both offensive and insensitive by the family and friends of those who died on 9/11 by jumping from the upper floors of the WTC towers. .

According to the IPT-ACS (Joiner, 2005), individuals who jumped from the WTC on 9/11 died because they elected to kill themselves. Joiner's rhetorical question (2005), "Did they die by suicide?" (p. 141) clearly suggests that he did not believe people who jumped from those towers were killed by al-Qaida. The answer Joiner provided to that rhetorical question contributes to the validity of our suggestion when he wrote: "According to the New York medical examiner, they did not. All September 11 deaths at the World Trade Center were classified as homicides" (Joiner, 2005, p. 141).

Furthermore, Joiner lamented that he could not find an easy way to explain deaths by jumping from the upper floors of the WTC as cases of suicides, because none of those who jumped from those towers on 9/11 survived to confirm that they had already felt the need of committing suicide and that the al-Qaida attacks simply provided the opportunity for the materialization of the suicidal behavior. Joiner wrote to express his disagreement with the New York examiner: "Had even one person survived from the upper floors from which people jumped, *perhaps a case for suicidal elements could be made, but no one from those floors survived*" (Joiner, 2005, p. 141, italics added). In this passage, Joiner clearly *lamented* that none of the individuals who jumped from those towers on 9/11survived to confirm the IPT-ACS and his disagreement with the New York examiner.

The question one may have in mind is this: How did Joiner arrive at the conclusion that people who jumped from the upper floors WTC towers committed suicide, rather than being murdered by the al-Qaida terrorists (as noted by the New York medical examiner)? The answer is that in Joiner's IPT-ACS jumping from those towers to death is the same as jumping from the San Francisco Golden Gate Bridge (Blaustein & Fleming, 2009). In both cases, the IPT-ACS proposes that the "ability to enact self-injury "(Joiner, 2005, pp. 46-93) and the "desire for death" (pp. 94-136) should explain both events. In order to fully understand the logic behind the IPT-ACS in explaining those deaths as suicides rather than homicides, it is important to remember that the phrase "desire for death" in the IPT-ACS is "composed of two psychological states—perceived burdensomeness and failed belongingness" (Joiner, 2005, p. 136).

According to Joiner (2005), despite the fact that the New York medical examiner classified all deaths on 9/11 as homicides, "still, the case can be made that these deaths [resulting from jumping from the upper floors of the WTC] *technically* were suicides, and if so, the current model [IPT-ACS] *should have something to say about them*" (p. 141, italics added). Joiner further stated:

"*The situation on the upper floors of the World Trade Center forced people to make horrible probability calculations…jumping meant a quick end to suffering and certain death; not jumping meant a minuscule chance of survival at the cost of near-certain death by more painful means. Who would cling* to this minuscule chance of survival? By the logic of the current model* [i.e., Joiner's theory], *there are two possible answers—those who wished to jump* [i.e., the desire for death] *but could not because they were unable to enact lethal self-injury* [i.e., the third required element in the theory], *and those whose desire to live was enormously strong—enough so to risk intensive pain and suffering in exchange for even the smallest chance to live*" (Joiner, 2005, p. 142, italics added).

In lay terms, what Joiner (2005) wanted to say to justify his disagreement with the New York medical examiner is that when people jump from high places with the intention to die by suicide (e.g., the Golden Gate Bridge, a four-story building, the WTC towers, etc.) they do so as a "quick end to suffering and certain death" (Joiner, 2005, p. 141). Before jumping, however, the suicidal individual must be able to enact lethal self-injury resulting from past habituation to pain and provocation this individual experienced in combination with the "desire for death" (i.e., the individual's perceived burdensomeness and thwarted belongingness). Therefore, in terms of the logic of Joiner's theory, on 9/11 jumping from the upper floors of the WTC was a case of suicidal behavior and not a case of homicide (as determined by the New York medical examiner), because the jumpers were ready to "enact lethal self-injury" as the result of their past (long before 9/11) habituation to pain and provocation in combination with their perceived burdensomeness and thwarted belongingness (i.e., their "desire for death") prior to the 9/11 al-Qaida terrorist attacks.

Inversely, in the logic of the IPT-ACS, individuals who decided not to jump from the upper floors of the WTC but elected to run down the stairs did that because of their *inability* to enact lethal self-injury (i.e., not enough habituation to pain and provocation prior to 9/11 to enact lethal self-injury) and they also did not perceived to be a burden to their family, friends, co-workers, etc., and did not have a sense of thwarted belongingness (i.e., they did not have a "desire for death"). Those "non-jumpers" sensed "a minuscule chance of survival [i.e., running down the stairs] at the cost of near-certain death by more painful means [i.e., jumping from those towers]" (Joiner, 2005, p. 141).

As noted above, Joiner lamented none of the jumpers from the upper floors of the WTC survived to confirm his theory of everything in this case (i.e., that the above three constructs were actually present before people jumped from those towers to their death; see Joiner, 2005, p. 141). Aside from this *unexpected* finding, Joiner did not report empirical evidence based on suicidal notes or psychological autopsies (see Shneidman, 2004) to support his offensive and distasteful conclusion that all the individuals who jumped from the WTC towers committed suicide, rather than being killed by the al-Qaida terrorist attacks on September 11, 2001. To our knowledge, the IPT-ACS is the only theory of suicide that has attempted to explain the death of those who jumped from the WTC towers in terms of suicidal behavior, instead of homicides resulting from the 9/11 attacks.

Summary

In the above discussion, we showed that the 9/11 al-Qaida terrorists did not die by suicide. These terrorists died because of their strong sense of martyrdom (as defined above). Joiner's (2005) theory of suicide, however, claims that those terrorists died by suicide, but without convincing empirical evidence to confirm the theory's assertion in this context. For example, an attempt to interview the family or friends of those terrorists would have been sufficient to show whether or not they actually felt to be a burden to their family (or friends), not meaningfully connected to others (thwarted belongingness), and had acquired the ability to enact lethal self-injury resulting from a long period of habituation to pain and provocation prior to the 9/11 attacks. In the absence of that evidence, the leaders of the al-Qaida terrorist organization are right when they claim a strong sense of martyrdom was the main factor leading to their death, and not suicide as predicted by Joiner's (2005) theory.

We predict the family and friends of those who jumped from the upper floors of the WTC on 9/11 will be surprised to know that a theory of suicide exists to explain that tragic event. The question is: Why do we need a theory of suicide to explain that event when the New York medical examiner determined that those who jumped from the towers were actually killed by the al-Qaida? The answer is that Joiner probably felt at the time he was writing his theory that if he could explain the deaths on 9/11 as suicides the three main constructs in the theory (i.e., thwarted belongingness, perceived burdensomeness, and the ability to enact lethal self-injury) would reach the apex of validity. We showed, however, that the theory has no validity in the present context and, more important, that the formulation of a theory of suicide to explain why those victims of 9/11 died by suicide and not by the al-Qaida terrorist attacks is a very insensitive argument for the family and friends of those victims.

If Joiner decides to revise his theory of everything, he should apologize to the family, friends and co-workers of those who jumped from the upper floors of the World Trade Center on 9/11 for going too far with the IPT-ACS.

VIOLATIONS OF FUNDAMENTAL ASSUMPTIONS IN THE IPT-ACS

The present section evaluates examples of qualitative and quantitative studies in support of the IPT-ACS. Paraphrasing Lester (2004b, p. 53), it is interesting to examine whether such studies did in fact test the IPT-ACS.

Qualitative Research

Although Joiner discussed suicidal ideation during the formulation of the theory (see Joiner, 2005, pp. 79-80), he suggested that an empirical test of the IPT-ACS would require a population of either suicide attempters or those who have died by suicide. Joiner was correct in avoiding the explanation of suicidal ideation with the IPT-ACS, for the following reason. As noted earlier, the theory requires the simultaneous presence of the above three constructs to explain suicidal behavior, and it was very clear to Joiner that at least one of those constructs (i.e., acquired ability to enact lethal self-injury) could not be empirically tested with individuals who only report suicidal ideation *without* attempt. Some researchers eager to test Joiner's theory, however, have not followed this specific underlined assumption in the IPT-ACS. Below are two examples of this point, with emphasis on qualitative research.

Brenner et al. (2008) adapted a diagram originally published by Joiner (2005) on page 138. In that diagram, Joiner did not include suicidal ideation as a critical element in the testing of his theory (see also Joiner & Van Orden, 2008, p. 82). Brenner et al (2008), however, added "suicidal ideation" (see p. 213) to Joiner's original diagram. In addition, Brenner et al. (2008) agreed that Joiner's theory only deals with "attempted and completed suicide" (p. 211). The sample in their study, however, included only individuals who reported suicidal ideation. Therefore, Brenner et al. (2008) tested Joiner's theory *with the wrong sample.*

We agree with Brenner et al. (2008) in that "qualitative research can be particularly useful in clarifying less-understood phenomena" (p. 213). Endless examples in the behavioral sciences confirm this observation (Creswell, 2009). The merit of a qualitative study, however, is recognized when its findings are derived from the best available reliable and valid methodological approach; otherwise, Kuhn's (1996) argument would apply: "quantitative predictions are preferable to qualitative ones" (p. 185), particularly in the case of testing a new theory that deals with a very complex issue in the behavioral sciences (e.g., suicidal behavior).

Aside from the fact that Brenner et al. (2008) used the wrong sample in the testing of Joiner's theory, a major methodological problem in their study is that verbal reports from participants do not appear to reflect, in several narratives, the core constructs in the IPT-ACS. For example, Brenner et al. (2008) summarized, on page 217, examples of participants' verbal reports to show the validity of the construct "acquired ability to enact lethal self-injury" (Joiner, 2005, pp. 46-93). These examples, however, do not provide convincing evidence that participants actually engaged in "lethal self-injury." In the IPT-ACS, an indirect measure of this construct is an assessment of either prior attempted suicide or death by suicide (see Joiner, 2005, p. 47). In the study by Brenner et al. (2008), however, none of the participants engaged in suicide attempt or death by suicide. Therefore, that construct was not empirically tested in that study.

Furthermore, in the Brenner et al. (2008) study, verbal reports suggesting "perceived burdensomeness" and "failed belongingness" also appear questionable in the context of confirming the validity of these two constructs in their study. For example, it does not appear that the verbal report "I was a good soldier. That's what I was good at" (Brenner, 2008, p. 218) reflects the "perceived burdensomeness" or the "thwarted belongingness" constructs in Joiner's theory. Finally, Brenner et al. (2008) pointed out that "transcripts were independently reviewed by another member of the team" (p. 215), but the levels of agreement (reliability) between the two raters were not documented.

Another qualitative study in violation of Joiner's main assumptions is the study by Anestis et al. (2009). Ironically, Joiner is the last author in this study and it appears that he contributed to the violation of his own stipulation regarding the need to avoid testing the IPT-ACS with cases of suicidal ideation. This study involved two cases of individuals on active-duty in the United States Air Force who neither attempted suicide nor died by suicide, but engaged in suicidal ideation. Anestis et al. (2009) pointed out, correctly, that Joiner's theory " posits that three variables—perceived burdensomeness, thwarted belongingness, and acquired capability for suicide—determine the risk for an individual engaging in a lethal suicide attempt" (p. 60). The problem with the interpretation of the results as confirmatory evidence for the test of the IPT-ACS is that in their study Anestis et al. (2009) did not include suicide attempters or individuals who died by suicide. Therefore, Anestis et al. (2009) also tested Joiner's theory with *the wrong sample* (i.e., individuals with only suicidal ideation).

Actually, Anestis et al. (2009) agreed with the above conclusion when they wrote: "because *neither client in these case studies actually attempted suicide*, it is impossible to definitively evaluate the degree to which the presence or absence of elevations in the variables of interest [the core constructs in Joiner's theory] was related to suicide behavior" (p. 73, italics added). As discussed above, this conclusion is particularly true in the case of the construct of "acquired ability to enact lethal self-injury," because if the individual does not engage in a suicide attempt or a completed suicide this would prevent an assessment of the degree of lethality assumed in that construct. In both cases reported by Anestis et al. (2009), this construct is clearly missed.

Despite the fact that Anestis et al. (2009) did not actually test Joiner's theory for reasons explained above, they concluded that their findings could 'help to understand the phenomenon of suicide in the military" (p. 60). Aside from methodological and experimental design problems with the Anestis et al. study, it is unethical in scientific research to suggest the generalization of findings to the population (e.g., suicide in the military population) in those cases when the sample (e.g., two cases in Anestis et al., 2009) selected from that population is very small and not representative of the population (Kazdin, 1992).

Quantitative Research

After the publication of the IPT-ACS in 2005, the number of quantitative studies substantially increased, relative to few qualitative studies (e.g., Conner, Britton, Sworts, & Joiner, 2007; Joiner, Van Orden, Witte, Selby, Ribeiro, Lewis, & Rudd, 2009; Nademin, Jobes, Pflanz, Jacoby, Ghahramnlou-Holloway, Campise, Joiner, Wagner, and Johnson; 2008; Van Orden, Witte, Gordon, Bender, & Joiner, 2008). Two of these studies are particularly relevant to illustrate the failure to test Joiner's theory main assumptions with quantitative research, namely, Nademin et al. (2008) and Van Orden et al. (2008).

Although qualitative studies have resulted in significant scientific contributions in the behavioral sciences, the general assumption is that quantitative research is preferable in the testing of theories (Creswell, 2009; Kuhn, 1996; Popper, 1935/1959). Quantitative research is not, however, synonymous with good or valid research. A case in point is the study by Nademin. et. al. (2008). The purpose of this study was to *"evaluate whether Joiner's theory could differentiate United States...Air Force (AF) personnel (n = 60) who died by suicide from a living active duty AF personnel sample (n = 122"* (p. 309, italics in the text). The experimental hypothesis in this study was: "greater acquired capability to suicide, greater sense of burdensomeness, and greater thwarted belongingness would differentiate those who died by suicide from living controls" (Nademin et al., 2008, p. 311).

Nademin et al. (2008) pointed out that in order "to more comprehensively assess for Joiner's three constructs" (p. 311), they developed the *Interpersonal-Psychological Survey* (IPS), which assesses all three constructs in Joiner's theory. In addition to the IPS, the study included the *Acquired Capability to Suicide Scale* –ACSS (for the assessment of the acquired capability to enact suicide in Joiner's theory), and the *Interpersonal Needs Questionnaire-INQ*, which is "designed to measure beliefs about how connected one feels to others (i.e., belongingness)" (Nademin et al., 2008, p. 313). Finally, the *Suicide Death Investigation Template* (SDIT) was employed with the AF sample who died by suicide. The SDIT was used in this study to collect "psychological and factual elements of the deceased's life. Information is based on retrospective chart data" (p. 312), but with the exclusion of information regarding Joiner's (2005) theoretical constructs.

Nademin et al. (2008) reported impressive results involving the psychometric properties of the IPS, ACSS, and the INQ, and they rejected the null hypothesis: "Findings indicate that the elements of Joiner's theory *collectively* differentiate between these two samples [i.e., living active duty AF personnel and AF individuals who died by suicide]" (p. 318, italics in the text). In the specific case of the IPS, Nademin et al. (2008) concluded that "overall IPS total score significantly differentiated between *those in the suicide group* and living group" (p. 318, italics added). Did Nademin et al. (2008) reject the null hypothesis? The answer is negative. Several methodological and experimental design deficiencies would confirm this conclusion, but only three are discussed below.

First, convenience sampling (Mitchell & Jolley, 1988) was used to select the living sample. As noted by Nademin et al. (2008), "passing AF personnel were asked to stop by a booth …to complete a brief series of surveys on suicide prevention. Each living active duty completed four measures…the IPS, ACSS, INQ, and demographic questionnaire" (p. 314). In addition, although the abstract of the study asserted that the sample who died by suicide was randomly selected from "postmortem investigatory files" (p. 309), the method section suggests that this sample was also selected through convenience sampling. Because of this methodological deficiency, one would conclude the samples in this study did not represent the population and, in the case of the living sample, asking participants to volunteer to complete questionnaires under that "passing by" condition probably resulted in an extremely biased sample (see Mitchell & Jolley, 1988, p. 304).

Second, despite the impressive psychometric results reported by Nademin et al. (2008), a critical issue is the absence of a measure of reliability across scales. Nademin et al. (2008) asserted that "inter-rater reliability of IPS data was high" (p. 316). It appears, however, that reliability measures with the living sample were not reported. Individuals in the living sample were asked to complete the scale during that "passing by" situation. Therefore, one would assume that the living sample did not return to repeat the same scale (this is inferred from the method section). If true, conclusions regarding the convergent and discriminant validity of measures (see Nademin et al., 2008, p. 317) are questionable, because a test that is not reliable is, by definition, an invalid test (Anastasi, 1988; Green, 1992).

Third, aside from the above relatively minor methodological problems, a major experimental design problem is that only the living sample completed the IPS, ACSS, and INQ. Therefore, in the specific case of the IPS, for example, it is essentially impossible to conclude that "overall IPS total score significantly differentiated between those in the suicide group and living group" (Nademin et al., 2008, p. 318) using x^2 statistics. The same conclusion applies in the case of Table 3 (p. 315), in which the authors reported the mean and the standard deviation for the suicide sample and the living sample in the case of their responses on the IPS. Given the obvious limitation of assessing the deceased, however, one would assume that the suicide sample did not receive the IPS (and neither of the other measures received by the living sample) in the study by Nademin et al. (2008).

Because Nademin et al. (2008) did not assess the suicide sample with the IPS, the conclusion that "overall IPS total score significantly differentiated between those in the suicide group and living group" (p. 318) is not supported with the present experimental design. The main critique in this context is not that a Type I (i.e., rejecting a null hypothesis when it is actually true) or a Type II error (i.e., failure to reject the null hypothesis when it is actually false) was committed in an attempt to reject the null hypothesis, but that Nademin et al. (2008) conducted the wrong experiment (i.e., they made a Type III error). In paraphrasing Smith and Sechrest (1992, p. 567), one would argue that faulty experimental designs and measurement prohibit meaningful interpretation of experimental results in the study by Nademin et al. (2008).

Furthermore, Nademin et al. (2008) found that, among the three core constructs in the IPT-ACS, "only capability to enact suicide, as measured by the IPS...emerged as a significant predictor of suicide risk. Therefore, this component [in the IPT-ACS] may serve *as a stand-alone risk factor for suicide within the sample*" (pp. 318-319, italics added). In terms of Joiner's (2005) theory, this conclusion is another argument in support of the fact that Nademin et al. (2008) did not test the IPT-ACS because none of Joiner's theoretical constructs can "stand-alone" in support of the explanatory power of the IPT-ACS; all three constructs must be present (see Martin et al., 2009, p. 111) to explain either suicide attempts or completed suicides.

In the case of the study by Van Orden et al. (2008), the authors reported three studies (henceforth, S1, S2, and S3) to test the "interpersonal-psychological theory of suicide behavior" (p. 72). Aside from the fact that the reported large (.43, Table 2) and medium (.16, Table 4) effect sizes do not correspond to Cohen's (1988) conventional definition of small (=. 20), medium (=. 50), and large (= .80) effect size, each of the studies reported by Van Orden et al. (2008) tested Joiner's theory *partially.*

If the IPT-ACS is used to predict who is at risk for engaging in a suicide attempt (*sa)* one must first demonstrate that, before the suicide attempt, the individual had experienced a "desire for death" (*dd*) in terms of perceived burdensomeness (*pb*) and thwarted belongingness (*tb*) as well as a long period of habituation and provocation (*hp*) leading (-) to the acquired ability to enact lethal self-injury (*ls*). Therefore: *sa = dd (Pb+tb) + (hp-ls).* If the theory is used to explain completed suicide, the equation is: (*cs*): *cs = dd (pb+tb) + (hp - ls) +/-(sa)*: completed suicide occurred because the individual had a desire for death, plus a period of habituation and provocation to averse situations leading to (-) lethal self-injury that might (+) or might not be (-) preceded by a history of suicide attempts (*sa*) as a crucial element developed during that period of *hp*.

The reason for the *+/-(sa)* in the second equation is to account for the so-called "out of the blue" suicidal behavior, in which the suicidal act occurs "without warning" (Heeringen, 2001c, p. 10) and in the absence of prior suicide attempts. In this pathway to suicide, individuals who failed at suicide would report that they "tried to kill themselves without any apparent suicidal ideation or preparation, other than in the very last moment. The urge to commit suicide was felt suddenly and irresistibly, and the impulse to kill oneself could not be controlled. These suicides can be precipitated by events which severely threatened self-esteem...[Examples of these events include] severe humiliation or financial loss" (Kerkhof & Arensman, 2001, p. 34).

In the first study (S1;Van Orden et al., 2008), the experimental question was: "who wants to die by suicide?" and the author tested this question with only *pb* and *tb*. In the second study (S2), the question was: "who can die by suicide?" and only *ls* was used to answer this question. The experimental question in third study (S3) was: "who is at greatest risk for suicide behavior?" and the answer was "those at greatest risk both desire suicide [i.e., the desire for death-*dd*- in Joiner's definition of this element in the theory] and are capable of suicidal behavior [i.e., the ability to enact lethal self-injury-*ls*]" (Van Orden et al., 2008, p. 78). As noted earlier, in Joiner's theory "the desire for death" (or *dd* in above equations) includes both perceived burdensomeness-*pb* and thwarted belongingness-*tb* (see Joiner, 2005, pp. 94-136). For reasons unexplained in S3, however, only perceived burdensomeness (*pb*) was assessed in combination with the acquired capability to enact lethal self-injury (*ls*). Therefore, Joiner's experimental demands to test the IPT-ACS were not followed in the case of S1, S2, and S3 because in each study all three constructs were not included in the same equation.

The first study (S1) in Van Orden et al. (2008) also failed to test Joiner's theory for two additional reasons. First, Joiner proposed a theory of suicide to explain why some individuals attempt suicide and why others kill themselves. This means that a test of the IPT-ACS cannot be performed with a non-clinical sample. In S1, however, the sample included healthy undergraduate students that were "... likely representative of the Florida State University... Psychology Department subject pool" (Van Orden et al., 2008, p. 73). This sample was not, however, representative of the clinical population the IPT-ACS proposes to include in either qualitative or quantitative research. This observation suggests that results derived from S1 could not be generalized to the clinical population the IPT-ACS requires for its empirical test (Haeffel, Thiessen, Campbell, Kaschak, & McNeil, 2009).

Second, the dependent variable in S1 was "suicidal ideation" (the independent variables were thwarted belongingness-*tb* and perceived burdensomeness-*pb*). This is not applicable in the IPT-ACS, because a critical element in the theory is the individual's ability to actually enact lethal self-injury (*ls*) and this construct cannot be tested if he individual only reports suicidal ideation without attempts.

It should be noted that Table 1 in S1 (Van Orden et al., 2008, p. 74) shows that the intercorrelations between depression (measured with the Beck Depression Inventory) and belongingness (r = .60), depression and burdensomeness (r =.71), and depression and suicidal ideation (r=.41; measured with the Beck Scale for Suicide Ideation) *were significant* at the p<.01. These results suggest that depression was a critical variable in the explanation of the results (e.g., participants who perceived burdensomeness or reported suicidal ideation were probably also depressed). As pointed out earlier, psychopathology plays a minimal role in the IPT-ACS; in S1, however, depression appeared to be a critical element.

In the case of S2 (Van Orden et al. (2008), the main findings were that past "suicidal attempts would predict acquired capability for suicide" (p. 77) and that the "lowest level of acquired capability was reported by individuals with no past suicide attempts" (p. 77). These findings, however, are not examples of a test of the IPT-ACS because the *ls* construct was assessed in the absence of the *pb* and *tb* constructs. Similar to findings with S1 relative to the impact of depression in the explanation of the results, S2 also showed strong significant intercorrelations (p<.05 to p<.01) among depression, suicidal ideation, burdensomeness (*pb*), and belongingness (*tb*). Similar points could be made in the case of S3, which tested the theory with only two constructs (three are required), the study includes only suicidal ideation (SI) as the main dependent measure (SI cannot be tested with Joiner's theory), and this study also found depression to be a significant predictor for suicidal ideation (see Van Orden et al., 2008, p. 79).

In the discussion of the three studies, Van Orden et al. (2008) pointed out that "the theory does not propose that thwarted belongingness [*tb*] and perceived burdensomeness [*pb*] are the *only paths to suicidal desire* but that their joint presence is likely to result in a highly pernicious form of suicidal desire" (p. 80, italics added). This is not, however, in accord with the main argument in Joiner's theory.

During the formulation of the IPT-ACS, Joiner (2005) emphasized that *the paths to suicidal acts* (i.e., suicide attempt and completed suicide) involve the presence of all three constructs in his theory (i.e., *ls, pb, and tb,* in the above equations). Joiner wrote: "Drawing on diverse literature, the case is made that people desire death when two fundamental needs *are frustrated* to the point of extinction; namely, the need to belong with or to connect to others [*tb*], and the need to feel effective with or to influence others [*pb*]. When both these needs are snuffed out, suicide becomes attractive *but not accessible without the ability for self-harm* [*ls*]" (Joiner, 2005, p. 47, italics added; see also Anestis et al., 2009, p. 60). This critical demand in the IPT-ACS to explain why people either attempt suicide or die by suicide is what both qualitative and quantitative research have missed in their efforts to test the IPT-ACS.

Summary

The central argument in the above discussion is that both quantitative and qualitative studies supporting Joiner's (2005) theory have violated two fundamental assumptions of the theory. First, the samples in these studies included individuals who expressed suicidal ideation, but in the absence of suicide attempt. Joiner's theory deals with suicide attempts and completed suicides. Second, in this theory all the psychological constructs (i.e., thwarted belongingness, perceived burdensomeness, and the ability to enact lethal self-injury) must be present for the theory to work appropriately in a given context. The studies reviewed in the above section, however, tested the theory *partially* (i.e., not all constructs were simultaneously examined).

PROBLEMS OF TESTABILITY/
FALSIFIABILITY WITH THE IPT-ACS

Popper (1935/1959) argued that "theories may be more, or less, severely testable, that is to say more, or less, easily falsifiable. The degree of their testability is of significance for the selection of theories" (p. 95). According to Popper (1935/1959), if the degree of testability or falsifiability required by the theory is beyond what would be expected by independent researchers (i.e., the theory cannot be shown to be either false or true with available methodological approaches and experimental designs), the theory, by definition, cannot be tested or falsified. A theory that fails to meet Popper's metric of testability or falsifiability, is not a scientific theory (Haeffel et al, 2009). We briefly discuss two areas to illustrate the problem of testability or falsifiability with the IPT-ACS, namely, that it is very difficult to exclude *traditional risks for suicidal behavior* in an empirical test of the IPT-ACS and that it is also impossible *to empirically test the construct of "vicarious habituation"* postulated by this theory.

The Exclusion of Traditional Risk Factors for Suicidal Behavior

As noted earlier, the IPT-ACS assumes that risks for attempted and completed suicide are primarily associated with the three constructs in the theory (e.g., perceived burdensomeness). If one does not experience all three constructs *together*, the risk for suicidal behavior would be near zero (see Anestis et al., 2009, p. 60). The complexity of suicidal behavior, however, requires a multifaceted approach that prohibits an emphasis on a limited number of risk factors (e.g., only three constructs in Joiner's theory). This observation probably explains why Anestis et al. (2009) concluded that "it is important to consider the potential impact of psychopathology on suicide risk" (p. 68) and not only Joiner's three constructs. These authors further revealed that in the specific case of symptoms of depression, such symptoms would have to be considered in an empirical study "as covariates in order to ascertain the degree to which the three components [constructs] of Joiner's theory account for variability in suicide risk *beyond the influence of depression*" (p. 68, italics added).

Anestis et al. (2009) were on the right track in their conclusion because, despite the possibility that Joiner's three constructs might very well play key roles in the explanation of suicide acts, traditional risk factors associated with attempted and completed suicide would have to be considered in order to appropriately deal with the complexity of explaining this problem (Lester, 2004b). In other words, an unbiased and independent test of the IPT-ACS would be practically impossible to conduct in the absence of the traditional risks for suicidal behavior generally considered in the suicidology literature (Harris & Barraclough, 1997; Jamison, 1999; Krysinska & Martin, 2009; Lester, 2004b; Simon & Hales, 2006).

Examples of well-researched traditional risks for suicide (some of them cited in Joiner's theory) include: unemployment and low income; marital problems; family history of suicide; early parental separation; genetic inheritance; some psychological constructs such as hopelessness, anger, frustration, hostility; the desire to escape from aversive situations; a manipulative behavior to gain attention from others (particularly during a suicide attempt); the individual's irrationality; an innate releasing (ethological) mechanism; a cultural or religious act; and psychopathology, plus an additional number of endless risks (Jamison, 1999; Lester, 2004b; Lieb, Bronisch, Hofler, Schreier, & Wittchen, 2005; Krysinska & Martin, 2009; Qin, Agerbo, & Mortensen, 2003; Papakostas, Petersen, Pava, Masson, Worthington, Alpert, Fava, & Niuerenberg, 2003; Shneidman, 1996; Shneidman, 1996; Simon & Hales, 2006).

In the specific case of psychopathology as a risk factor for suicidal behavior, Joiner's agreement regarding that "95 percent of people who die by suicide experienced a mental disorder at the time of death" (Joiner, 2005, p. 191) is welcome recognition that talking about suicide *without* seriously talking about psychiatric illness in most suicidal cases would bring current theorists of suicide back to Emile Durkheim's time (i.e., 1897).

The Impossibility of Empirically Testing the "Vicarious Habituation" Construct

The second area in Joiner's theory that is problematic to test is the obvious problem of designing an experiment to empirically investigate the role of the "vicarious habituation" construct in the IPT-ACS. As noted earlier, Joiner (2005) argues one does not need to *directly* experience pain, violence, and injury leading to one's ability to engage in lethal self-injury resulting in a suicide attempt or completed suicide; one could habituate to these situations by *observing* others displaying those painful and provocative events.

For example, Joiner used the construct of "vicarious habituation" to explain why the suicide rate among physicians is higher than the rate among other professionals (Joiner, 2005). Physicians, relative to psychologists, sociologists, physicists, lawyers, and other professionals, have the opportunity to "frequently *observe* the consequences of pain, violence, and injury [in others], and they gain specialized knowledge about lethal agents, dosing, methods of death, and so forth" (Joiner, 2005, p. 73, italics added); this is what habituates them to pain and provocation leading to their acquired ability to enact lethal self-injury (*ls*). Physicians who kill themselves, however, would do so when the other two constructs in the IPT-ACS are also present (i.e., perceived burdensomeness-*pb*- and thwarted belongingness-*tb*).

Despite the potential merit of the "vicarious habituation" construct in the IPT-ACS, this construct has not yet been empirically evaluated in a test of the theory. The reason for this observation is that it would be practically impossible or not "easily falsifiable" (Popper, 1035/1959, p. 95) to experimentally test the following hypothesis: *Subjects in the experimental group will engage in lethal self-injury (ls) behaviors resulting either in suicide attempt or death by suicide after being exposed to the observation of repeated "vicarious habituation" in others, whereas subjects in the control group will not acquire the capability to enact lethal self-injury after watching others engaging in non-painful and non-provocative situations.*

In that hypothesis, one would assume that subjects in the experimental groups would report "perceived burdensomeness" (*pb*) and "thwarted belongingness" (*tb*), whereas the control group would not report these constructs. With the help of minimal methodological criteria (e.g., random selection of subjects from a population, random assignment of subjects to the experimental versus the control group, and use of reliable and valid measures; Kazdin, 1992), the null hypothesis might very well be rejected: *only* subjects in the experimental group, indeed, acquired the ability to enact lethal self-injury during the programming of the "vicarious habitation" construct. In other words, the investigator would happily report that with the help of the "vicarious habituation" construct the experimental sample was able to habituate "to stimuli that previously would have been highly aversive, with respect to both fear and physiological response" (Anestis et al., 2009, p. 61).

Despite the above potential finding, it would be very difficult to actually design a *non-hypothetical* experiment to test the role of the "vicarious habituation" construct in the IPT-ACS. This is an obstacle that would clearly prevent investigators from applying Popper's (1935/1959) criterion of falsifiability or testability, for two reasons. First, participants in the experimental group instructed to observe others "habituating" (independent variable) to pain and provocation would have to either attempt suicide or die by suicide (dependent variables) in order to reject the null hypothesis. Second, institutional review boards would never allow researchers to conduct an experiment to test that "vicarious habituation" construct with humans. Therefore, even if the "vicarious habituation" construct has merit in the IPT-ACS it would be more than impossible to test or falsify this construct in the IPT-ACS (Popper, 1935/1959).

It should be noted that anecdotal reports (Coleman, 1987; Jamison, 1999) and scientific evidence (de Leo & Heller, 2008; Gould, Jamieson, & Romer, 2003; Lester, 1990; Lester, 2004b) have documented that imitation (or learning through observation of other people's behavior; Bandura, 1969, 1977; Paniagua & Baer, 1981) could be a "process by which one suicide becomes a compelling model for successive suicides" (Williams & Pollock, 2001, p. 89). The process of "vicarious habituation" in Joiner's theory (2005), however, appears to differ from the concepts of "imitation" or "observational learning" in Bandura's social learning theory (e.g., Bandura, 1977; Paniagua & Baer, 1981).

In the case of the modeling or imitation of suicide behavior in others, the rate of suicide at any given time might increase among certain individuals in the population following "a widely published suicide story" (Williams & Pollock, 2001, p. 90). This is known as the "media effect on completed suicide" (Williams & Pollock, p. 90; see also Lester, 1990, p. 67, and Jamison 1999, p 143-153). In the imitation of suicidal behavior, however, a single occurrence of the suicide event is sufficient for others to imitate the event in the absence of other particularities leading to lethal self-injury (*ls*).

For example, the media might report that John Doe jumped from the San Francisco Golden Bridge Gate to his death; the next day Jane Doe imitates this event by committing suicide from the same bridge. Assuming that John Doe prompted the suicidal behavior of Jane Doe, this is an example of "modeling suicidal behavior" (Williams & Pollock, 2001, p. 89) and not an example of "vicarious habituation" because, prior to Jane Doe's decision to kill herself, she did not observe (through the process of "vicarious habituation") John Doe engaging in a long period of habituation and provocation (*hp*) involving lethal self-injury (*ls*, e.g., prior suicide attempts), violence, pain, and other particularities associated with the "vicarious habitation" construct in the IPT-ACS. It should be noted that Joiner (2005) did not cite Bandura's social learning theory during the discussion of the "vicarious habituation" construct in the formulation of the IPT-ACS (e.g., see Joiner, 2005, pp. 72-73). This observation suggests that perhaps Joiner realized that "vicarious habituation" is not the same as "imitation" or "observational learning" in Bandura's social learning theory.

Summary

The overall conclusion in the preceding discussion is that the problem of confirming the three main constructs of the IPT-ACS with the *exclusion of traditional risk factors* leading to either suicide attempts or completed suicides (particularly when a clinical population is required to test the theory), and the impossibility of experimentally testing or falsifying the "vicarious habituation" construct in the IPT-ACS, are two elements that would prevent independent researchers from testing or falsifying the IPT-ACS (see Haeffel et al., 2009, p. 570).

CONCLUSION

The multifactorial nature of suicide is what makes this phenomenon extremely difficult to understand. This observation explains why so many theories of suicide have been proposed to explain why people either attempt suicide or die by suicide (Lester, 2004a; Lester, 2004b; Lester, 2008). Any attempt to formulate a theory of suicide that is reduced to an extremely limited number of risk factors (e.g., above three constructs in the ITP-ACS) has the potential to make a minimal contribution to the field of suicidology. In addition, a theory should include a review of the existing literature that might conflict with the main argument of the theory. The original formulation of the IPT-ACS in 2005 clearly did not meet this standard, and post-2005 publications involving this theory have not yet considered this suggestion (e.g., Joiner et al., 2009; Joiner, Van Orden, Witte, & Rudd, 2009).

Furthermore, if key elements of a theory cannot be falsified (Popper, 1935/1959) with existing methodological and experimental design procedures the theory is more speculative than scientific (Haeffel et al., 2009). In the case of Joiner's (2005) theory of suicide, it is clearly impossible to test the three key psychological constructs in the theory in the absence of the potential impact of traditional risk factors generally associated with suicidal acts. This is particularly true in the case of psychopathology, which has been found in prior studies a strong predictor for increasing rates of attempts or completed suicide among mentally ill individuals (Harris & Barraclough, 1997; Moscicki, 2001).

It is important, however, to note that Joiner's approach has some merit if it is viewed not as a theory of suicide but as a framework that could unify the exemplars (Kuhn, 1996) of diverse theories of suicide. This suggestion is actually implicit in the original formulation of the IPT-ACS, when Joiner (2005) described how key psychological states in the theory of suicide proposed by Baumeister (1990), Beck (1976), Durkheim (1897/1951), Linehan (1993), and Shneidman (1996) are very similar to the three constructs in the IPT-ACS. For example, in the case of Beck's theory (Beck, 1976; Beck, Brown, Berchick, & Stewart, 1990) rather than talking about "*hopelessness*" in isolation a better context for this construct would be to associate it with psychological states in the IPT-ACS (e.g., perceived burdensomeness and thwarted belongingness). In paraphrasing Joiner (2005, p. 39), in this example the question "hopelessness about what?" would lead to the answer "about feeling a burden to others and not connected to a valued group."

FOOTNOTES

[1] In the suicidology context, the Standard Mortality Ration (SMR) measures the risk of suicidal behavior in a given mental disorder in comparison with the expected rate in the general population (with a SMR of 1). The SMR "is calculated for each [mental] disorder by dividing observed mortality by expected mortality" (Simon, 2006, p. 18). In the case of the study by Harris and Barraclough (1997), "they compared observed number of suicides in individuals with mental disorders with those expected in the general population" (Simon, 2006, p. 18).

[2] The Population Attributable Risk (PAR) is "the proportional reduction in average disease risk over a specific time interval what would be achieved by eliminating the exposure(s) of interest from the population while distributions of other risk factors in the population remain unchanged. This also can be interpreted as the proportion of disease cases over a specific time that would be prevented following elimination of the exposures, assuming the exposures are causal" (Rockhill, Newman, & Weinberg, 1998, p. 15). The population attributable risk is generally written as $PAR = \{Pr\ (D) - Pr\ (D/\bar{E}\}\ /Pr\ (D)$, "where Pr (D) is the probability of disease in the population, which may have some exposed, E, and some unexposed, $\bar{E}$, individuals, and Pr $(D/\bar{E})$ is the hypothetical probability of disease in the same population but with all exposure estimated" (Benichou, 2001, p. 195). For several derivations of the above equation, see Benichou (2001, pp. 195-196), Hahn, Eaker, Barker, Teutch, Sosniak, and Krieger (1995, p. 492), and Rockhill et al, (1998, p. 17, Table 1). In the field of suicidology, the PAR estimates the "percentage reduction in suicide rates that would occur if there was a causal association between the risk factor and suicide, and the risk factor was eliminated from the population" (Krysinska & Martin, 2009, p. 549).

[3]These are only two of many examples of radical and wrong interpretations of the Qur'an by al-Qaida terrorists (Moghaddam, 2006; Paniagua, 2005). For example, in *Surah* 22.39 and 22.40, respectively, the Qur'an gives permission to fight, but in self-defense (Paniagua, 2007). Al-Qaida terrorists on 9/11, however, believed that the Qur'an gives permission to fight under any situation if the main goal is to kill the "unbelievers" (Paniagua, 2007). In addition, in *Surah* 2.256 the Qur'an states that "there is not compulsion in religion," and this means that Islam is only one of many religions in the world (Habeck, 2006).

REFERENCES

Anastasi, A. (1988). *Psychological testing.* New York: McMillan.

Anestis, M.D., Bryan, C. J., Cornette, M.M., & Joiner, T.E. (2009). Understanding suicide behavior in the Military: An evaluation of Joiner's interpersonal-psychological theory of suicide behavior in two case studies of active duty post-deployers. *Journal of Mental Health Counseling, 31, pp. 60-75.*

Bandura, A. (1969). <u>Principles of behavior modification</u>. New York: Holt, Rinehart, & Winston.

Bandura, A. (1977). *Social learning theory.* Englewood Cliffs, NJ: Prentice-Hall.

Baumeister, R. F. (1990). Suicide as escape from self. *Psychological Review*, 97, 90-113.

Beck, A. T. (1976). Cognitive therapy and the emotional disorders. New York: International University Press.

Beck, A. T. (1996). Beyond belief: A theory of modes, personality, and psychopathology. In P. Salkovskis (Ed.), *Frontiers of cognitive therapy* (pp. 1-25). New York: Guilford Press.

Beck, A. T., Brown, G., & Steer, R. A. (1989). Prediction of eventual suicide in psychiatric inpatients by clinical ratings of hopelessness. *Journal of Consulting and Clinical Psychology*, 57, 309-310

Beck, A. T., Brown, G., Berchick, R. J., & Stewart, B. L. (1990). Relationship between hopelessness and ultimate suicide: A replication with psychiatric outpatients. *American Journal of Psychiatry*, 147, 190-195.

Benichou, J. (2001). A review of adjusted estimators of attributable risk. *Statistical Methods in Medical Research*, 10, 195-2126.

Blaustein, M., & Fleming, A. (2009). Suicide from the Golden Gate bridge. *American Journal of Psychiatry*, 166, 1111-1116.

Brenner, L. A., Gutierrez, P. M., Cornetter, M.M., Bethauser, L. M., Bahraini, N., & Staves, P. J. (2008). A qualitative study of potential suicide risk factors in returning combat Veterans. *Journal of Mental Health Counseling*, 30, pp. 211-225.

Brown, G. K., Jeglic, E., Henriques, G., & Beck, A. T. (2006). Cognitive therapy, cognition, and suicidal behavior. In T. E. Ellis (Ed.), *Cognition and suicide: Theory, research, and therapy* (pp. 53-74). Washington, DC: American Psychological Association.

Brown, M. Z. (2006). Linehan's theory of suicidal behavior: Theory, research, and dialectical Behavior therapy. In T. E. Ellis (Ed.), *Cognition and suicide: Theory, research, and therapy* (pp. 591-117). Washington, DC: American Psychological Association.

Center for Disease Control and Prevention (2009a). *Twenty leading causes of death highlighting suicide among persons ages 10 years and older, United States, 2006.* Retrieved from http://www. cdc.gov/violenceprevention/suicide/statistics/leading_causes. html, December 21, 2009.

Center for Disease Control and Prevention (2009b). *Trends in suicide rates among males, by age group, United States, 1991-2006.* Retrieved on December 21, 2009 from http://www.cdc.gov/ violenceprevention/suicide/statistics/trends03.html.

Chou, Y. J., Huang, N., Lee, C. H., Tsai, S. L., Tsay, J. H., Chen, L. S., & Chou, P. (2003). Suicides after the 1999 Taiwan earthquake. *International Journal of Epidemiology*, 32, 1007-1014.

Coleman, L. (1987). *Suicide clusters.* Boston, Massachusetts: Faber & Faber.

Conner, K. R., Britton, P. C., Sworts, L. M., & Joiner, T. E. (2007). Suicide attempts among individuals with opiate dependence:

The critical role of belonging. *Addictive Behavior*, 32, 1395-1404.

Creswell, J. W. (2009). *Research design: Qualitative, quantitative, and mixed methods approaches* (3rd Ed.). Thousand Oaks, CA: Sage Publications.

de Leo, D., & Heller, T. (2008). Social modeling in the transmission of suicidality. Crisis, 29, 11-19.

Ditzler, T. F. (2004). Malevolent minds: The teleology of terrorism. In F. M. Moghaddam & A. J. Marsella (Eds.), *Understanding terrorism:* Psychological *Psychosocial roots, consequences, and interventions* (pp. 187-206). Washington, DC: American Psychological Association.

Durkheim, E. (1951). *Suicide: A study of sociology* (J. A. Spaulding & G Simpson, Trans.). New York: The Free Press. (Original work published 1897).

Ellis, T. E. (Ed.). (2007). Cognition *and suicide: Theory, research,* and therapy. Washington, DC: Amer4ican Psychological Association.

Emerson, S. (2006). *Jihad incorporated: A guide to militant Islam in the US*. Amherst, N.Y.: Prometheus Books.

Gold, L. H. (2006). Suicide and gender. In R. I. Simon, & R. E. Hales (Eds.), *Textbook of suicide assessment and management* (pp. 77-106). Washington, DC: American Psychiatric Press.

Goldney, R. D. (2001). Ethology and suicide process. In K. V. Heeringen (Ed.), *Understanding suicidal behavior: The suicidal process approach to research, treatment and prevention* (pp. 120-135). New York: John Wiley & Sons.

Gould, M., Jamieson, P., & Romer, D. (2003). Media contagious and suicide among the young. *American Behavioral Scientist*, 46, 1269-1284.

Habeck, M. (2006). *Knowing the enemy: Jihadist ideology and the war on terror*. New Heaven, CT: Yale University Press.

Haeffel, G. J., Thiessen, E. D., Campbell, M. W., Kaschak, & McNeil, N. M. (2009). Theory, not cultural context, will advance American psychology. *American Psychologist*, 64, 570-571.

Hahn, R. A., Eaker, E., Barker, N., Teutsch, S. M., Sosniak, W., & Krieger, N. (1995). Poverty and death in the United States-1973 and 1991. *Epidemiology*, 6, 490-497.

Harris, C. E, & Barraclough, B. (1997). Suicide as an outcome for mental disorders. *British Journal of Psychiatry*, 170, 2005-228.

Heeringen, K. V. (Ed.). (2001a). *Understanding suicidal behavior: The suicidal process approach to* research, *treatment and prevention.* New York: John Wiley & Sons.

Heeringen, K. V. (2001b). Towards a psychobiological model of the suicidal process. In K.V, Heeringen (Ed.), *Understanding suicidal behavior: The suicidal process approach to research, treatment and prevention* (pp. 136-159). New York: John Wiley & Sons.

Heeringen, K. V. (2001c). The suicidal process and related concepts. In K. V. Heeringen (Ed.), *Understanding suicidal behavior: The suicidal process approach to research, treatment and prevention* (pp. 3-14). New York: John Wiley & Sons.

Jamison, K. R. (1999). *Night falls fast: Understanding suicide.* New York: Vintage Books.

Jobes, L. A., & Nelson, K. N. (2006). Shneidman's contributions to the understanding of suicidal thinking. In T. E. Ellis (Ed.), *Cognition and suicide: Theory, research, and therapy* (pp. 29-49). Washington, DC: American Psychological Association.

Joiner, T. E. (2005). *Why people die by suicide.* Cambridge, MA: Harvard University Press.

Joiner, T. E., & Van Orden, K. A. (2008). The interpersonal-psychological theory of suicide behavior indicates specific and crucial psychotherapeutic targets. *International Journal of Cognitive Therapy*, 1, 80-89.

Joiner, T. E., Van Orden, K. A., Witte, T. K., Selby, E. A., Ribeiro, J. D., Lewis, R., & Rudd, M. D. (2009). Main predictions of interpersonal-psychological theory of suicidal behavior: Empirical tests in two samples of young adults. *Journal of Abnormal Psychology*, 118, 634-646.

Joiner, T. E., Van Orden, K. A., Witte, T. K., & Rudd, M. D. (2009). *The interpersonal theory of suicide: Guidance for working with suicidal clients*. Washington, DC: American Psychological Association.

Kazdin, A. E. (Ed.). (1992). *Methodological issues & strategies in clinical research*. Washington, DC: American Psychological Association.

Kerkhof, J.F.M, & Arensman, E. (2001. Pathways to suicide: The epidemiology of suicidal process. In K. V. Heeringen (Ed.), *Understanding suicidal behavior: The suicidal process approach to research, treatment and prevention* (pp. 15-39). New York: John Wiley & Sons.

Krysinska, K., & Martin, G. (2009). The struggle to prevent and evaluate: Application of popular attributable risk and preventive fraction to suicide prevention research. *Suicide and Life-Threatening Behavior*, 39, 548-557.

Kuhn, T. S. (1996). The *structure of scientific revolutions (3rd ed.)*. Illinois: University of Chicago Press.

Leenaars, A. A. (2008). Suicide: A cross-cultural theory. In Frederick T. L. Leong & Mark M. Leach (Eds.), *Suicide among racial and ethnic minority groups* (pp. 13-37). New York: Taylor and Francis Group.

Lester, D. (1990). *Understanding and preventing suicide*. Springfield, Illinois: Charles C. Thomas Publisher.

Lester, D. (1994). A comparison of 15 theories of suicide. *Suicide and Life-Threatening Behavior*, 24, 80-88.

Lester, D. (2004a). A comparison of fifteen theories of suicide. *Suicide & Life-Threatening Behavior*, 24, pp. 80-88.

Lester, D. (2004b). *Thinking about suicide: Perspectives on suicide*. New York: Nova Science Publishers.

Lester, D. (2008). Theories of suicide. In Frederick T. L. Leong & Mark M. Leach (Eds.), *Suicide among racial and ethnic minority groups* (pp. 39-53). . New York Taylor and Francis Group.

Lieb, R., Bronisch, T., Hofler, M., Schreier, & Wittchen, H. U. (2005). Maternal suicidality and risk of suicidality in offspring: Findings from a community study. *American Journal of Psychiatry, 162,* 1665-1671.

Linehan, M.M. (1993). *Cognitive-behavioral treatment of borderline personality disorder.* New York: Gilford Press.

Mahan, S., & Griset, P. L. (Eds.). (2008). Terrorist tactics around the globe. In S. Mahan & P. L. Griset (Eds.), *Terrorism in perspective* (pp. 129-144). Thousand Oaks, CA: Sage Publications.

Mann, J. J., Waternaux, C., Haas, G. L., & Malone, K. (1999). Toward a clinical model of suicide behavior in psychiatric patients. *American Journal of Psychiatry,* 156, 181-189.

Martin, J., Ghahramanlou-Holloway, L M., Lou, K., & Tucciarone, P. (2009). A comparative review of U.S. Military and civilian suicide behavior: Implications for OEF/OIF suicide prevention efforts. *Journal of Mental Health Counseling,* 31, pp. 101-118.

Marsella, A. J. (2004). Reflections on international terrorism: Issues, concepts, and Directions. In F. M. Moghaddam & A. J. Marsella (Eds.), *Understanding terrorism:* Psychological *Psychosocial roots, consequences, and interventions* (pp. 11-47). Washington, DC: American Psychological Association.

Mitchell, M., & Jolley, J. (1988). *Research design explained.* New York: Holt, Rinehart and Winston.

Moghaddam, F. M. (2006). *From the terrorists' point of view: What they experience and why they come to destroy.* New York: Praeger.

Moghaddam, F. M. & Marsella, A. J. (Eds.). (2004).*Understanding terrorism: Psychological roots, consequences, and interventions.* Washington, DC: American Psychological Association.

Moscicki, E. K. (2001). Epidemiology of completed and attempted suicide: Toward a framework for prevention. *Clinical Neuroscience Research*, 1, 310-323.

Murray, H. A. (1938). *Explorations in personality.* New York: Oxford University Press.

Nademin, E., Jobes, D. A., Pflanz, S. E. , Jacoby, A. M., Ghahramanlou-Holloway, M., Campise, R., Joiner, T, Wagner, B. M., & Johnson, L. (2008). An investigation of interpersonal- psychological variables in Air Force suicides: A controlled-comparison study. *Archives of Suicide Research*, 12, 309-326.

Paniagua, F. A. (2005). Some thoughts on the "staircase to terrorism." *American Psychologist*, 60, 1038-1039.

Paniagua, F. A. (2006) From the Terrorists' Point of View. [Review of the book *From the Terrorists' Point of View: What They Experience and Why They Come to Destroy*]. *Journal of Homeland Security and Emergency Management*, 3(4), Article 10.

Paniagua, F. A. (2007) Understanding Second-Variant Jihadists. [Review of the book *Knowing the Enemy: Jihadist Ideology* and *the War on Terror*]. *Contemporary Psychology* 52, article 149, pp. 1-10.

Paniagua, F. A. (2009). Treating suicidal soldiers. *Houston Chronicle*, February 9, p. B8.

Paniagua, F. A. (2010a). Some commends to further improve the DoDSER. *Military Medicine*, 175, 80-81

Paniagua, F. A. (2010b). Suicide among racial and ethnic minority groups [Review of the book "Suicide among racial and ethnic minority groups" edited by Frederick T. L. Leong and Mark M. Leach]. *Cultural Diversity & Ethnic Minority Psychology*, 16, 297-298.

Paniagua, F. A., & Baer, D. M. (1981). A procedural analysis of the symbolic forms of behavior therapy. *Behaviorism*, 9, 171-205.

Papakostas, G. I., Petersen, T., Pava, J., Masson, E., Worthington, J. J., Alpert, J. E., Fava, M., & Nierenberg, A. A. (2003). Hopelessness and suicidal ideation in outpatients with treatment-resistant depression: Prevalence and impact on treatment outcome. *Journal of Nervous and Mental Disease, 191,* 444-449.

Popper, K. (1959). *The logic of scientific discovery* (K. Popper, J. Freed, & L. Freed, Trans.). New York: Rutledge. (Originally published 1935).

Qin, P., Agerbo, E., & Mortensen, P. B. (2003). Suicide risk in relation to socioeconomic, demographic, psychiatric, and familial factors: A national registered-based study of all suicides in Denmark, 1981-1997. *American Journal of Psychiatry, 160,* 765-772.

Rockhill, B., Newman, & Weinberg, C. (1998). Use and misuse of population attributable fractions. *American Journal of Public Health, 88,* 15-19

Roy, A., Nielsen, D., Rylander, G., Sarchiapone, M., & Segal, N. (1999). Genetics of suicide in depression. *Journal of Clinical Psychiatry, 60,* 12-17

Roy, A., Rylander, G., & Sarchiapone, M. (1997). Genetics of suicide: Family studies and Molecular genetics. *Annals of the New York Academy of Sciences, 836,* 135-157.

Rudd, M. D. (2000). The suicidal mode: A cognitive-behavioral model of suicidality. *Suicide and Life Threatening Behavior*, 30, 18-33.

Rudd, M. D., Joiner, T. E., & Rajab, M. H. (2001). *Treating suicide behavior: An effective time limited approach*. New York: Guilford Press.

Shneidman, E S. (1987). A psychological approach to suicide. In G. R. VandenBos & B. K. Bryant (Eds.), Cataclysms, crises, and catastrophes: Psychology in action (pp. 147-183. Washington, DC: American Psychological Association.

Shneidman, E. S. (1996). *The suicidal mind*. New York: Oxford University Press.

Shneidman, E. S. (2004). *Autopsy of a suicidal mind*. New York: Oxford University Press.

Simon, R. I. (2006). Suicide risk: Assessing the unpredictable. In Robert I. Simon & Robert E. Hales (Eds.), *Textbook of suicide assessment and management* (pp. 1-32). Washington, DC: American Psychiatric Press.

Simon, R. I., & Robert E. Hales (Eds.), (2006). *Textbook of suicide assessment and management* Washington, DC: American Psychiatric Press.

Smith, B., & Sechrest, L. (1992). Treatment of aptitude X treatment interactions. In A. E. Kazdin (Ed.), *Methodological issues & strategies in clinical research.* (pp. 557-584). Washington, DC: American Psychological Association.

Stern, J. (2008). The ultimate organization: Networks, franchises, and freelances. In S. Mahan & P. L. Griset (Eds.), *Terrorism in perspective* (pp. 154-179). Thousand Oaks, CA: Sage Publications.

Tomassini, C., Juel, K., Holm, N. V., Skytthe, A., & Christensen, K. (2003). Risk of suicide in twins: 51 year follow up study. *British Medical Journal*, 327, 373-374.

Van Orden, K. A., Witte, T. K., Gordon, K. H., Bender, T. W., & Joiner, T. E. (2008). Suicide desire and capability for suicide: Test of the interpersonal-psychological theory of suicidal behavior among adults. *Journal of Consulting and Clinical Psychology*, 76, 72-83.

Williams, J. M., & Pollock, L. R. (2001). Psychological aspects of the suicidal process. In K. V. Heeringen (Ed.), *Understanding suicidal behavior: The suicidal process approach to research, treatment and prevention* (pp. 77-93). New York: John Wiley & Sons.

Wingate, L. R., Burns, A. B., Gordon, K. H., Perez, M., Walker, R. L., Williams, F. M., & Joiner, T E. (2006). Suicide and positive cognitions: Positive psychology applied to the understanding and treatment of suicidal behavior. In T. E. Ellis (Ed.), *Cognition and suicide: Theory, research, and therapy* (pp. 261-283). Washington, DC: American Psychological Association.

World Health Organization (2009*). Suicide prevention.* Retrieved September 28, 2009 from http://www.who.int/mental_health/prevention/suicide/suicideprevent/en/

Yampey, N. (1967). Consideraciones epidemiologicas sobre el suicidio en Buenos Aires. *Acta Psiquiatrica y Psicologica de America Latina,* 13, 39-44.

ABOUT THE AUTHORS

Dr. Freddy A. Paniagua is retired Tenured Professor from the University of Texas Medical Branch (UTMB) at Galveston, and currently Adjunct Professor at UTMB. Dr. Paniagua was (2009-2010) associated with the Faculty Research Participation Program at the U.S. Public Health Command (USPHC) administered by the Oak Ridge Institute for Science and Education (ORISE), and assigned to the Behavioral and Social Health Outcomes Program (BSHOP). Dr. Sandra A. Black was a senior epidemiologist associated with BSHOP at the time this book was produced. Dr. Shayne M. Gallaway is a senior epidemiologist at BSHOP. Ms. Michelle A. Coombs is also associated with BSHOP and is a doctoral clinical psychology student, Fielding Graduate University, Santa Barbara, California.